Revise
AS

D1336245

Physics

Graham Booth & David Brodie

Contents

Chapter 3 Waves, imaging and information

Chapter 4 Waves, particles and the Universe

AS/A2 Level Physics courses

AS and A2

All Physics A Level courses are in two parts, with three separate units or modules in each part. Most students will start by studying the AS (Advanced Subsidiary) course. Some will then go on to study the second part of the A Level course, called the A2. It is also possible to study the full A Level course, both AS and A2, in any order.

How will you be tested?

Assessment units

For AS Physics, you will be tested by three assessment units. For the full A Level in Physics, you will take a further three units. AS Physics forms 50% of the assessment weighting for the full A Level.

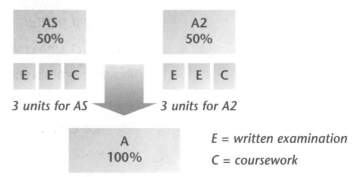

Units that are assessed by written examination can normally be taken in either January or June. Coursework units can be completed for assessment only in June. It is also possible to study the whole course before taking any of the unit tests. There is a lot of flexibility about when exams can be taken and the diagram below shows just some of the ways that the assessment units may be taken for AS and A Level Physics.

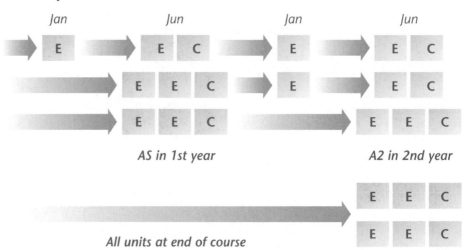

If you are disappointed with a unit result then you can re-sit the unit examination. The higher mark counts.

External assessment

For external assessment, written examinations are set and marked by teachers and lecturers from across the country, and overseen by the awarding bodies.

Internal assessment

Internal assessment is based on work that you do over a period of time, and is marked by your own teachers or lecturers. Some or all of your internally assessed work is based on practical and investigative skills and communication of your work.

What will you need to be able to do?

You will be tested by 'assessment objectives'. These are the skills and abilities that you should have acquired by studying the course. The assessment objectives for AS Physics are shown below.

Knowledge and understanding of science and how science works

- recall and understanding of scientific knowledge
- selection, organisation and communication of information in different forms

Application of knowledge and understanding of science and how science works

- analysis and evaluation of knowledge and processes
- applying knowledge and processes to unfamiliar situations
- assessing the validity, reliability and credibility of scientific information

How science works

- use ethical, safe and skilful practical methods
- select appropriate methods and make valid observations and measurements with appropriate precision and accuracy
- record and communicate observations and conclusions
- evaluate methods, results and conclusions

The awarding bodies and their specifications

The awarding bodies in England, Wales and Northern Ireland are AQA, CCEA, Edexcel, OCR and WJEC/CBAC. These awarding bodies oversee the development of specifications and examinations as well as marking and moderation.

AQA A Physics

ASSESSMENT UNIT	TOPIC, SECTION OR MODULE	CHAPTER REFERENCE	STUDIED	REVISED	PRACTICE QUESTIONS
1. Particles, quantum phenomena and electricity	Particles and radiation	4.1, 4.2, 4.3			
	Electromagnetic radiation and quantum phenomena	3.1, 4.3			
	Current electricity	2.1, 2.2, 2.3			
2. Mechanics, materials and waves	Mechanics	1.1, 1.2, 1.3, 1.4, 1.5, 1.6			
	Materials	1.7			
	Waves	3.1, 3.2, 3.3, 3.4			
3. Investigative and practical skills, internally assessed					

AS assessment analysis

Unit 1	1 h 15 min test	40%
Unit 2	1 h 15 min test	40%
Unit 3	Internally assessed practical skills and investigative skills	20%

AQA B Physics

ASSESSMENT UNIT	TOPIC, SECTION OR MODULE	CHAPTER REFERENCE	STUDIED	REVISED	PRACTICE QUESTIONS
1. Harmony and structure in the Universe	The world of music	3.1, 3.2, 3.3, 3.4			
	From quarks to quasars	4.1, 4.2, 4.3, 4.4			
2. Physics keeps us going	Moving people, people moving	1.1, 1.2, 1.3, 1.4, 1.5, 1.6			
	Energy and the environment	1.4, 2.1, 2.2, 2.3			
3. Investigative and practical skills, internally assessed					

Examination analysis

Unit 1	1 h 15 min test	40%
Unit 2	1 h 15 min test	40%
Unit 3	Internally assessed practical skills and investigative skills	20%

CCEA Physics

ASSESSMENT UNIT	TOPIC, SECTION OR MODULE	CHAPTER REFERENCE	STUDIED	REVISED	PRACTICE QUESTIONS
1. Forces, energy and electricity	Mechanics	1.1, 1.2, 1.3, 1.4, 1.5, 1.6, 1.7			
	Electricity	2.1, 2.2, 2.3			
2. Waves, photons and Medical Physics	Waves and medical applications	3.1, 3.2, 3.3, 3.4			
	Quantum Physics	4.3, 4.4			
3. Practical techniques					

AS assessment analysis

Unit 1	1 h 30 min test	37%
Unit 2	1 h 30 min test	37%
Unit 3	Internally assessed practical techniques	26%

Edexcel Physics, concept approach

ASSESSMENT UNIT	TOPIC, SECTION OR MODULE	CHAPTER REFERENCE	STUDIED	REVISED	PRACTICE QUESTIONS
1. Physics on the go	Mechanics	1.1, 1.2, 1.3, 1.4, 1.5, 1.6			
	Materials	1.5, 1.7			
2. Physics at work	Waves	3.1, 3.2, 3.3, 3.4			
	Direct current electricity	2.1, 2.2, 2.3			
	The nature of light	3.1, 4.3			
3. Exploring Physics					

AS assessment analysis

Unit 1	1 h 20 min test	*40%*
Unit 2	1 h 20 min test	*40%*
Unit 3	Internally assessed practical experiment based on a visit or case study	*20%*

Edexcel Physics, context approach

ASSESSMENT UNIT	TOPIC, SECTION OR MODULE	CHAPTER REFERENCE	STUDIED	REVISED	PRACTICE QUESTIONS
1. Physics on the go	Higher, faster, stronger	1.1, 1.2, 1.3, 1.4, 1.5, 1.6			
	Good enough to eat	1.5, 1.7			
	Spare part surgery	1.7			
2. Physics at work	The sound of music	3.1, 3.2, 3.3, 3.4			
	Technology in space	2.1, 2.2, 2.3, 3.1, 4.3			
	Digging up the past	2.2, 3.1, 3.2, 3.3			
3. Exploring Physics					

AS assessment analysis

Unit 1	1 h 20 min test	40%
Unit 2	1 h 20 min test	40%
Unit 3	Internally assessed practical experiment based on a visit or case study	20%

OCR Physics A

ASSESSMENT UNIT	TOPIC, SECTION OR MODULE	CHAPTER REFERENCE	STUDIED	REVISED	PRACTICE QUESTIONS
1. Mechanics	Motion	1.1, 1.2			
	Forces in action	1.3, 1.5, 1.6			
	Work and energy	1.4, 1.7			
2. Electrons, waves and photons	Electric current, resistance and direct current circuits	2.1, 2.2, 2.3			
	Waves	3.1, 3.2, 3.3, 3.4			
	Quantum Physics	4.3			
3. Practical skills in Physics					

AS assessment analysis

Unit 1	1 h test	30%
Unit 2	1 h 45 min test	50%
Unit 3	Internally marked practical tasks	20%

OCR Physics B

ASSESSMENT UNIT	TOPIC, SECTION OR MODULE	CHAPTER REFERENCE	STUDIED	REVISED	PRACTICE QUESTIONS
1. Physics in action	Communication	2.1, 2.2, 2.3, 3.1, 3.4			
	Designer materials	1.7, 2.1, 2.2			
2. Understanding processes, and experimentation and data handling	Waves and quantum behaviour	3.1, 3.2, 3.3, 4.3			
	Space, time and motion	1.1, 1.2, 1.3, 1.4, 1.5			
3. Physics in practice					

AS assessment analysis

Unit 1	1 h test	30%
Unit 2	1 h 45 min test	50%
Unit 3	Internally marked practical tasks	20%

WJEC/CBAC Physics

ASSESSMENT UNIT	TOPIC, SECTION OR MODULE	CHAPTER REFERENCE	STUDIED	REVISED	PRACTICE QUESTIONS
1. Motion, energy and charge	Basic physics, kinematics and energy concepts	1.1, 1.2, 1.3, 1.4, 1.5, 1.6			
	Conduction of electricity, resistance and direct current circuits	2.1, 2.2, 2.3			
2. Waves and particles	Waves	3.1, 3.2, 3.3			
	Photons; matter, forces and the Universe; and using radiation to investigate stars	4.1, 4.2, 4.3, 4.4			
3. Practical Physics					

AS assessment analysis

Unit 1	1 h 15 min test	40%
Unit 2	1 h 15 min test	40%
Unit 3	Internally assessed experimental tasks	20%

Different types of questions in AS examinations

In AS Level Physics unit examinations, there may be combinations of short-answer questions and structured questions requiring both short answers and more extended answers. There may be some free-response and open-ended questions. There may also be some questions with a multiple-choice or 'objective' format.

Short-answer questions

A short-answer question may test recall or it may test understanding by requiring you to undertake a short, one-stage calculation. Short-answer questions normally have space for the answers printed on the question paper. Here are some examples (the answers are shown in blue):

What is the relationship between electric current and charge flow?

Current = rate of flow of charge.

The current passing in a heater is 6 A when it operates from 240 V mains. Calculate the power of the heating element.

$P = I \times V = 6\,A \times 240\,V = 1440\,W$

Which of the following is the correct unit of acceleration?

    A $m\,s^{-1}$
    B $s\,m^{-1}$
    C $m\,s^{-2}$
    D $m^2\,s^{-1}$

    C $m\,s^{-2}$

Structured questions

Structured questions are in several parts. The parts are usually about a common context and they often become progressively more difficult and more demanding as you work your way through the question. They may start with simple recall, then test understanding of a familiar or an unfamiliar situation. The most difficult part of a structured question is usually at the end, where the candidate is sometimes asked to suggest a reason for a particular phenomenon or social implication.

When answering structured questions, do not feel that you have to complete one question before starting the next. The further you are into a question, the more difficult the marks are to obtain. If you run out of ideas, go on to the next question. Five minutes spent on the beginning of that question are likely to be much more fruitful than the same time spent racking your brains trying to think of an explanation for an unfamiliar phenomenon.

Here is an example of a structured question that becomes progressively more demanding.

(a) A car speeds up from $20\,m\,s^{-1}$ to $50\,m\,s^{-1}$ in 15 s.

    Calculate the acceleration of the car.

      acceleration = increase in velocity ÷ time taken
                = $30\,m\,s^{-1} \div 15\,s = 2\,m\,s^{-1}$

(b) The total mass of the car and contents is 950 kg.

Calculate the size of the unbalanced force required to cause this acceleration.

force = mass × acceleration
     = 950 kg × 2 m s⁻² = 1900 N

(c) Suggest why the size of the driving force acting on the car needs to be greater than the answer to (b).

The driving force also has to do work to overcome the resistive forces, e.g. air resistance and rolling resistance.

Extended answers

In AS Level Physics, questions requiring more extended answers will usually form part of structured questions. They will normally appear at the end of structured questions and be characterised by having at least three marks (and often more, typically five) allocated to the answers as well as several lines (up to ten) of answer space. These questions are also used to assess your abilities to communicate ideas and put together a logical argument.

The correct answers to extended questions are less well-defined than those for short-answer questions. Examiners may have a list of points for which credit is awarded up to the maximum for the question, or they may first of all judge the quality of your response as poor, satisfactory or good before allocating it a mark within a range that corresponds to that quality.

As an example of a question that requires an extended answer, a structured question on the use of solar energy could end with the following:

Suggest why very few buildings make use of solar energy in this country compared to countries in southern Europe.   [5]

Points that the examiners might look for include:

• the energy from the Sun is unreliable due to cloud cover
• the intensity of the Sun's radiation is less in this country than in southern Europe due to the Earth's curvature
• more energy is absorbed by the atmosphere as the radiation has a greater depth of atmosphere to travel through
• fossil fuels are in abundant supply and relatively cheap
• the capital cost is high, giving a long payback time
• photo-voltaic cells have a low efficiency
• the energy is difficult to store for the times when it is needed the most

Full marks would be awarded for an argument that puts forward three or four of these points in a clear and logical way.

Free-response questions

AS Level Physics papers may or may not make use of free-response and open-ended questions. These types of questions allow you to choose the context and to develop your own ideas. Examples could include 'Describe a laboratory method of determining g, the value of free-fall acceleration' and 'Outline the evidence that suggests that light has a wave-like behaviour'. When answering these types of questions it is important to plan your response and present your answer in a logical order.

Exam technique

Advanced Subsidiary (AS) Physics builds from grade C in GCSE Science and GCSE Additional Science (combined) or GCSE Physics. This Study Guide has been written so that you will be able to tackle AS Physics from a GCSE Science background.

You should not need to search for important Physics from GCSE Science because this has been included where needed in each chapter. If you have not studied Science for some time, you should still be able to learn AS Physics using this text alone.

What are examiners looking for?

Examiners use instructions to help you to decide the length and depth of your answer. If a question does not seem to make sense, you may have misread it – read it again!

State, define or list

This requires a short, concise answer, often recall of material that can be learnt by rote.

Explain, describe or discuss

Some reasoning or reference to theory is required, depending on the context.

Outline

This implies a short response, almost a list of sentences or bullet points.

Predict or deduce

You are not expected to answer by recall but by making a connection between pieces of information.

Suggest

You are expected to apply your general knowledge to a 'novel' situation, one which you have not directly studied during the AS Physics course.

Calculate

This is used when a numerical answer is required. You should always use units in quantities and use significant figures with care. Look to see how many significant figures have been used for quantities in the question and give your answer to this degree of precision. If the question uses 3 (sig figs), then give your answer to 3 (sig figs) also.

Some dos and don'ts

Dos

Do answer the question

- No credit can be given for good Physics that is irrelevant to the question.

Do use the mark allocation to guide how much you write

- Two marks are awarded for two valid points – writing more will rarely gain more credit and could mean wasted time or even contradicting earlier valid points.

Do use diagrams, equations and tables in your responses

- Even in 'essay-type' questions, these offer an excellent way of communicating Physics. It is worth your while to practise drawing good clear and labelled sketches. Normally, you should do this without use of rulers or other drawing instruments.

Do write legibly

- An examiner cannot give marks if the answer cannot be read.

Do write using correct spelling and grammar. Structure longer essays carefully

- Marks are now awarded for the quality of your language in exams.

Don'ts

Don't fill up any blank space on a paper

- In structured questions, the number of dotted lines should guide the length of your answer.
- If you write too much, you waste time and may not finish the exam paper. You also risk contradicting yourself.

Don't write out the question again

- This wastes time. The marks are for the answer!

Don't contradict yourself

- The examiner cannot be expected to choose which answer is intended. You could lose a hard-earned mark.

Don't spend too much time on a part that you find difficult

- You may not have enough time to complete the exam. You can always return to a difficult calculation if you have time at the end of the exam.

What grade do you want?

Everyone would like to improve their grades but you will only manage this with a lot of hard work and determination. You should have a fair idea of your natural ability and likely grade in Physics and the hints below offer advice on improving that grade.

For a Grade A

You will need to be a very good all-rounder.

- You must go into every exam knowing the work extremely well.
- You must be able to apply your knowledge to new, unfamiliar situations.
- You need to have practised many, many exam questions so that you are ready for the type of question that will appear.

The exams test all areas of the specification and any weaknesses in your Physics will be found out. There must be no holes in your knowledge and understanding. For a Grade A, you must be competent in all areas.

For a Grade C

You must have a reasonable grasp of Physics but you may have weaknesses in several areas and you will be unsure of some of the reasons for the Physics.

- Many Grade C candidates are just as good at answering questions as the Grade A students but holes and weaknesses often show up in just some topics.
- To improve, you will need to master your weaknesses and you must prepare thoroughly for the exam. You must become a better all-rounder.

For a Grade E

You cannot afford to miss the easy marks. Even if you find Physics difficult to understand and would be happy with a Grade E, there are plenty of questions in which you can gain marks.

- You must memorise all definitions.
- You must practise exam questions to give yourself confidence that you do know some Physics. In exams, answer the parts of questions that you know first. You must not waste time on the difficult parts. You can always go back to these later.
- The areas of Physics that you find most difficult are going to be hard to score on in exams. Even in the difficult questions, there are still marks to be gained. Show your working in calculations because credit is given for a sound method. You can always gain some marks if you get part of the way towards the solution.

What marks do you need?

As a rough guide, you will need to score an average of 40% for a Grade E, 60% for a Grade C and 80% for a Grade A:

average	80%	70%	60%	50%	40%
grade	A	B	C	D	E

Essential mathematics

This section describes some of the mathematical techniques that are needed in studying AS Level Physics.

Quantities and units

Physical quantities are described by the appropriate words or symbols, for example the symbol R is used as shorthand for the value of a *resistance*. The quantity that the word or symbol represents has both a numerical value and a unit, e.g. 10.5 Ω. When writing data in a table or plotting a graph, only the numerical values are entered or plotted. For this reason headings used in tables and labels on graph axes are always written as (physical quantity)/(unit), where the slash represents division. When a physical quantity is divided by its unit, the result is the numerical value of the quantity.

Resistance is an example of a **derived** quantity and the ohm is a derived unit. This means that they are defined in terms of other quantities and units. All derived quantities and units can be expressed in terms of the seven **base** quantities and units of the SI, or International System of Units.

The quantities, their units and symbols are shown in the table. The candela is not used in AS or A2 Level Physics.

Quantity	Unit	Symbol
length	metre	m
mass	kilogram	kg
time	second	s
electric current	ampere	A
temperature difference	kelvin	K
amount of substance	mole	mol
luminous intensity	candela	cd

Equations

Physical quantities and homogeneous equations

The equations that you use in Physics are relationships between physical quantities. The value of a physical quantity includes both the numerical value and the unit it is measured in.

An equation must be **homogeneous**. That is, the units on each side of the equation must be the same.

For example, the equation:

4 cats + 5 dogs = 9 camels

is nonsense because it is not homogeneous.

Likewise:

4 A + 5 V = 9 Ω

is not homogeneous, and makes no sense.

However, the equations:

4 cats + 5 cats = 9 cats

and:

4 A + 5 A = 9 A

are homogeneous and correct.

Checking homogeneity in an equation is useful for:

• finding the units of a constant such as resistivity

- checking the possible correctness of an equation; if the units on each side are the same, the equation may be correct, but if they are different it is definitely wrong.

When to include units after values

A physicist will often write the equation:

$$4\,A \quad + \quad 5\,A \quad = \quad 9\,A$$

as:

$$4 \quad + \quad 5 \quad = \quad 9$$
$$\text{total current} \quad = \quad 9\,A$$

That is, it is permissible to leave out the unit in working, providing that the unit is given with the answer.

The equals sign

The = sign is at the heart of Mathematics, and it is a good habit never to use it incorrectly. In Physics especially, **the = sign is telling us that the two quantities either side are physically identical**. They may not look the same in the equation, but in the observable world we would not be able to tell the difference between one and the other. We cannot, for example, tell the difference between (4 + 5) cats and 9 cats however long we stare at them. This is a rule that is so obvious that people often forget it.

We can tell the difference between (4 + 5 + 1) cats and 9 cats (if they keep still for long enough).

(4 + 5 + 1) cats ≠ 9 cats

The ≠ sign means **not** equal. An equals sign in this 'equation' would not be telling the truth.

We cannot tell the difference between (4 + 5 + 1) cats and (9 + 1) cats.

(4 + 5 + 1) cats = (9 + 1) cats

The equals sign is telling the truth.

This reveals another rule that is also quite simple, but people often find it hard because they forget that = signs can only ever be used to tell the truth about indistinguishability. The new rule is that **you cannot change one side of an equation without changing the other in the same way.**

Rearranging equations

Often equations need to be used in a different form from that in which they are given or remembered. The equation needs to be rearranged. The rules for rearranging are:

- both sides of the equation must be changed in exactly the same way
- add and subtract before multiplying and dividing, and finally deal with roots and powers.

For example, suppose that you know the equation $v^2 = u^2 + 2as$ (and you know that the = sign here is telling the truth), and you want to work out u.

First, subtract 2as from both sides.

$$v^2 - 2as = u^2$$

(Check that you feel happy that this = sign is being honest. It will be if you have done the same thing to both sides of the original equation.)

Second, 'square root' both sides of the equation.

$$\sqrt{(v^2 - 2as)} = u$$

The hard part is not changing the equation (provided you remember the rules) but knowing what changes to make to get from the original form to the one you want.

It's not rocket science, but it does take practice.

For simple equations such as V = IR there is an alternative method, using the 'magic triangle'.

Write the equation in to the triangle. Cover up the quantity you want to work out. The pattern in the triangle tells you the equation you want.

The alternative method is to apply the rules as above, and change both sides of the equation in the same way. So if you divide both sides by R, you get V/R = I.

For equations such as these, it makes sense to use whichever method works best for you.

Drawing graphs

Graphs have a number of uses in Physics:

- they give an immediate, visual display of the relationship between physical quantities
- they enable the values of quantities to be determined
- they can be used to support or disprove a hypothesis about the relationship between variables.

When plotting a graph, it is important to remember that values determined by experiment are not exact. Every measurement has a certain level of accuracy and a certain level of precision.

An accurate measurement is one that agrees with the 'true' value. (Of course, the only way to find out a value of a physical quantity is to measure it, or to calculate it from other measured values. So the 'true' value is an ideal, and may be impossible to know.)

A precise measurement is one that agrees closely with repeated measurements. Precise measurements generally allow values to be given with confidence to more decimal places.

Perfect accuracy and perfect precision are fundamentally impossible. For these reasons, having plotted experimental values on a grid, the graph line is drawn as the best straight line or smooth curve that represents the points. Where there are 'anomalous' results, i.e. points that do not fit the straight line or curve, these should always be checked. If in doubt, ignore them, but do add a note in your experimental work to explain why you have ignored them and suggest how any anomalous results could have arisen.

Graphs, gradients and rates of change

Quantities such as velocity, acceleration and power are defined in terms of a **rate of change** of another quantity with time. This rate of change can be determined by calculating the gradient of an appropriate graph. For example, *velocity* is the *rate of change of displacement with time*. Its value is represented by the gradient of a displacement–time graph. Different techniques are used to determine the gradient of a straight line and a smooth curve. For a straight line:

- determine the value of $\varnothing y$, the change in the value of the quantity plotted on the y-axis, using the whole of the straight line part of the graph

A common error when determining the gradient of a graph is to work it out using the gridlines only, without reference to the scales on each axis.

- determine the corresponding value of Δx
- calculate the gradient as $\Delta y \div \Delta x$

For a smooth curve, the gradient is calculated by first drawing a tangent to the curve and then using the above method to determine the gradient of the tangent. To draw a tangent to a curve:

- mark the point on the curve where the gradient is to be determined
- use a pair of compasses to mark in two points on the curve, close to and equidistant from the point where the gradient is to be determined
- join these points with a ruler and extend the line beyond each point.

These techniques are illustrated in the diagrams below.

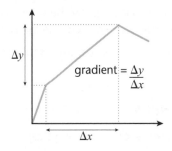

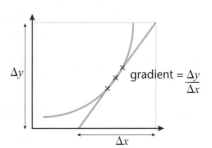

This calculation provides the value of the gradient of the middle section of the graph only. The other two sections have different gradients.

This method provides the value of the gradient at the middle of the three red marks. The graph has a continuously changing gradient that starts small and increases.

The area under a curve

The area between a graph line and the horizontal axis, often referred to as the 'area under the graph' can also yield useful information. This is the case when the product of the quantities plotted on the axes represents another physical quantity.

For example, on a *speed–time* graph this area represents the distance travelled. In the case of a straight line, the area can be calculated as that of the appropriate geometric figure. Where the graph line is curved, then the method of 'counting squares' is used.

- Count the number of complete squares between the graph line and the horizontal axis.
- Fractions of squares are counted as '1' if half the square or more is under the line, otherwise '0'.
- To work out the physical significance of each square, multiply together the quantities represented by one grid division on each axis.

These techniques are illustrated in the diagrams below:

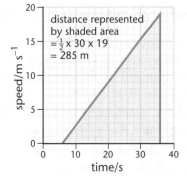

distance represented by shaded area
$= \frac{1}{2} \times 30 \times 19$
$= 285$ m

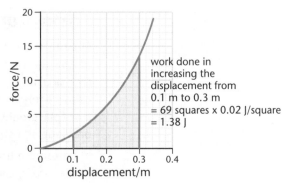

work done in increasing the displacement from 0.1 m to 0.3 m
= 69 squares × 0.02 J/square
= 1.38 J

Equations from graphs

By plotting the values of two variable quantities on a suitable graph, it may be possible to determine the relationship between the variables. This is straightforward when the graph is a straight line, since all straight line graphs have an equation of the form $y = mx + c$, where m is the gradient of the graph and c is the value of y when x is zero, i.e. the intercept on the y-axis. The relationship between the variables is determined by finding the values of m and c.

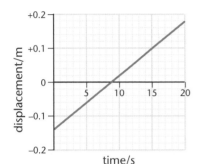

gradient = 0.32 ÷ 20
    = 0.016 m s^{-1}
intercept on y-axis = −0.14

equation is s = 0.016t − 0.14

The straight line graph on the left shows how the displacement, s, of an object varies with time, t.

The equation that describes this motion is:

s = 0.016t − 0.14

(Note that no units are shown in the equation, for simplicity. The unit for displacement, s, is the metre and the unit for time, t, is the second. The physical value 0.016 has unit metre per second, and the physical value 0.14 is measured in metres.)

There is an important and special case of the equation $y = mx + c$. This is the case for which c is zero, so that $y = mx$. The straight line of this graph now passes through the graph's origin.

m is the gradient of the graph and is constant (since the graph is a single straight line). Note that different symbols can be used for the constant gradient in place of m. It is quite common, for example, to use k. That gives $y = kx$.

In Physics, we are often seeking simple patterns, and this is about as simple as patterns can be. If $y = mx$ or $y = kx$, whatever changes happen to x then y always changes by the same proportion.

Trigonometry and Pythagoras

In the right-angled triangle shown here, the sides are labelled o (opposite), a (adjacent) and h (hypotenuse). The relationships between the size of the angle θ and the lengths of these sides are shown on the left.

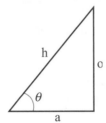

- sin θ = o/h
- cos θ = a/h
- tan θ = o/a

Pythagoras' theorem gives the relationship between the sides of a right-angled triangle: $h^2 = a^2 + o^2$

Multiples

For AS Level Physics, you are expected to be familiar with the following multiples of units:

Name	Multiple	Symbol
pico-	10^{-12}	p
nano-	10^{-9}	n
micro-	10^{-6}	µ
milli-	10^{-3}	m
kilo-	10^{3}	k
mega-	10^{6}	M
giga-	10^{9}	G

For example, the symbol MHz means 1×10^6 Hz and mN means 1×10^{-3} N.

Four steps to successful revision

Step 1: Understand

- Mark up the text if necessary – underline, highlight and make notes.
- Re-read each paragraph slowly.

GO TO STEP 2

Step 2: Summarise

- Now make your own revision note summary:
  What is the main idea, theme or concept to be learnt?
  What are the main points? How does the logic develop?
  Ask questions: Why? How? What next?
- Use bullet points, mind maps, patterned notes.
- Link ideas with mnemonics, mind maps, crazy stories.
- Note the title and date of the revision notes
  (e.g. Physics: Electricity, 3rd March).
- Organise your notes carefully and keep them in a file.

This is now in **short-term memory**. You will forget 80% of it if you do not go to Step 3.
GO TO STEP 3, but first take a 10 minute break.

Step 3: Memorise

- Take 25 minute learning 'bites' with 5 minute breaks.
- After each 5 minute break test yourself:
  Cover the original revision note summary
  Write down the main points
  Speak out loud (record on tape)
  Tell someone else
  Repeat many times.

The material is well on its way to **long-term memory**.
You will forget 40% if you do not do Step 4. **GO TO STEP 4**

Step 4: Track/Review

- Create a Revision Diary (one A4 page per day).
- Make a revision plan for the topic, e.g. 1 day later, 1 week later,
  1 month later.
- Record your revision in your Revision Diary, e.g.
  Physics: Electricity, 3rd March 25 minutes
  Physics: Electricity, 5th March 15 minutes
  Physics: Electricity, 3rd April 15 minutes
  ... revisit each topic at monthly intervals.

Chapter 1

Force, motion and energy

The following topics are covered in this chapter:

- *Vectors*
- *Representing and predicting motion*
- *Force and acceleration*
- *Energy and work*

- *Vehicles in motion*
- *More effects of forces*
- *Force and materials*

1.1 Vectors

After studying this section you should be able to:

- *recognise and describe forces*
- *distinguish between a vector and a scalar quantity*
- *add together two vectors and subtract one vector from another*
- *split a vector into two parts at right angles to each other*

LEARNING SUMMARY

Force as a vector quantity

AQA A	2	Edexcel context	1
AQA B	2	OCR A	1
CCEA	1	OCR B	2
Edexcel concept	1	WJEC	1

In this context 'normal' means 'at right angles to the surface'.

The first bullet point here is emphasising that all forces are caused by objects and they act on other objects.

Forces can be of different types: **gravitational** forces are caused by and affect objects with mass, **friction** forces oppose relative motion and the force pushing up on you at the moment, the **normal contact** force, is due to compression of the material that you are sat on.

Forces also have things in common:

- all forces can be described as **object A pulls/pushes object B**
- all forces can be represented in both size and direction by an arrow on a diagram.

Here are some examples.

A gravitational force

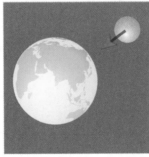

the Earth pulls
the Moon

A friction force

the ground pushes
the shoe

A normal contact force

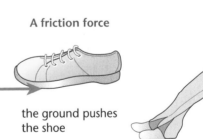

the chair pushes
the person

Physical quantities that have direction as well as size are called **vectors.** Quantities with size only are **scalars**. Some examples are given in the table.

When representing a vector quantity on a diagram, an arrow is always used to show its direction.

Vectors	Scalars
force	mass
velocity	speed
acceleration	length
displacement	distance
field strength	energy

Weight as a vector acting at the centre of gravity

AQA A	2	Edexcel context	1
AQA B	2	OCR A	1
CCEA	1	OCR B	2
Edexcel concept	1	WJEC	1

Gravitational forces act at a distance with no contact being necessary. These forces are always drawn as if they act at the **centre of gravity** of the object.

The centre of gravity of a traffic cone, a rubber ring and a person

The gravitational force acting on an object due to the planet or moon whose surface it is on is known as its weight. The centre of gravity of the object is the point at which its weight can be considered to act.

More vectors

AQA A	2	Edexcel context	1
AQA B	2	OCR A	1
CCEA	1	OCR B	2
Edexcel concept	1	WJEC	1

Information about the **speed** of an object only states how fast it is moving; the **velocity** also gives the direction. Speed is a scalar quantity while velocity is a vector quantity. For an object moving along a straight line, positive (+) and negative (−) are usually used to indicate movement in opposite directions.

As long as it is moving, the **distance** travelled by the object is increasing, but its **displacement** can increase or decrease and, like velocity, can have both positive and negative values.

> The fixed point is usually the starting point of the motion or the object's rest position.

Displacement is a vector quantity; it specifies both the distance and direction of an object measured from a fixed point.

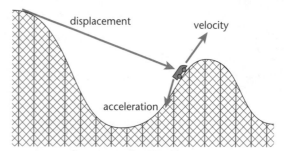

The diagram below shows a distance–time graph and a displacement–time graph for one particular journey.

> Check that you understand why distance has only positive values, but displacement has both positive and negative values.

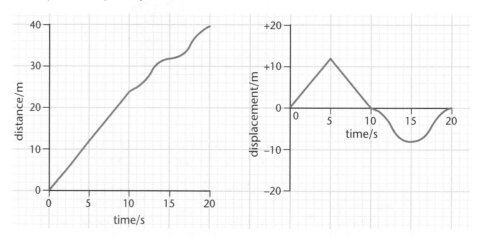

The same symbols are used for speed and velocity. When an object is changing speed/velocity, u is often used for the initial speed/velocity, with v being used for the final speed/velocity. The symbol c is usually reserved for the speed of light.

Information from the graphs can be used to calculate the **average speed** and **average velocity** over any time period, using the relationships:

average speed	= distance travelled ÷ time taken	$v = \Delta d \div \Delta t$
average velocity	= displacement ÷ time taken	$v = \Delta s \div \Delta t$

These show that over the 20 s time period, the average speed is $2\,\text{m s}^{-1}$ and the average velocity is 0.

Adding scalars and vectors

AQA A	2	Edexcel context	1
AQA B	2	OCR A	1
CCEA	1	OCR B	2
Edexcel concept	1	WJEC	1

To add together two scalar quantities the normal rules of arithmetic apply, for example, $2\,\text{kg} + 3\,\text{kg} = 5\,\text{kg}$ and no other answers are possible. When adding vector quantities, both the size and direction have to be taken into account.

What is the sum of a 2 N force and a 3 N force acting on the same object? The answer could be any value between 1 N and 5 N, depending on the directions involved.

> The sum, or resultant, of two vectors such as two forces acting on a single object is the single vector that could replace the two and have the same effect.

There are three steps to finding the sum of two vectors. These are illustrated by working out the sum of a 2 N force acting up the page and a 3 N force acting from left to right, both forces acting on the same object.

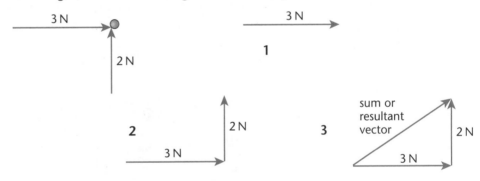

The order in which you draw the vectors does not matter; you should arrive at the same answer whichever one you start with.

- Draw an arrow that represents one of the vectors in both size and direction.
- Starting where this arrow finishes, draw an arrow that represents the second vector in size and direction.
- The sum, or resultant of the two vectors is represented (in both size and direction) by the single arrow drawn from the **start** of the first arrow to the **finish** of the second arrow.

In the example given, the size of the resultant force is 3.6 N and the direction is at an angle of 34° to the 3 N force. These figures were obtained by scale drawing.

Although a scale drawing is often the quickest way of working out the resultant of two vectors, the size and direction can also be calculated. This is straightforward when the vectors act at right angles, but needs more complex mathematics in other cases.

In the worked example:

The notation 'tan⁻¹' means 'the angle whose tangent is'.

- the size of the resultant force can be calculated using Pythagoras' theorem as $\sqrt{(2^2 + 3^2)}$
- the angle between the resultant force and the 3 N force can be calculated using the definition of tangent as $\tan^{-1}(2 \div 3)$.

Checking that the resultant is zero

The drawing method described above can be used to find the sum of any number of vectors by drawing an arrow for each vector, starting each new arrow where the previous one finished. The resultant of the vectors is then represented by the single arrow that starts at the beginning of the first vector and ends where the last one finishes. If the vectors being added together form a closed figure, i.e. the last one finishes where the first one starts, it follows that the sum is zero. This is what you would expect to find when working out the resultant force at a point in a stable structure, for example.

For equilibrium at point X, the sum of the forces must be zero.

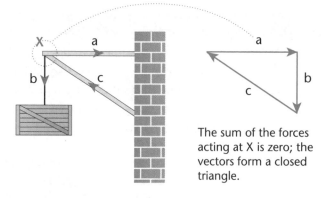

The sum of the forces acting at X is zero; the vectors form a closed triangle.

Splitting a vector in two

A vector has an effect in any direction except the one which is at right angles to it. Sometimes a vector has two independent effects which need to be isolated.

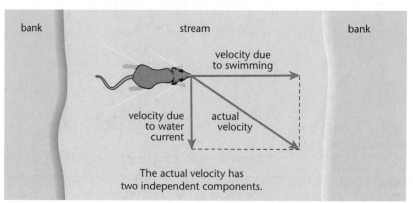

The actual velocity has two independent components.

Just as the combined effect of two vectors acting on a single object can be calculated, two separate effects of a single vector can be found by splitting the vector into two **components**. Provided that the directions of the two components are chosen to be at right angles, each one has no effect in the direction of the other so they are considered to act independently.

The process of splitting a vector into two components is known as **resolving** or **resolution of** the vector.

The diagram below shows the tension (T) in a cable holding a radio mast in place. The force is pulling the mast both vertically downwards and horizontally to the left.

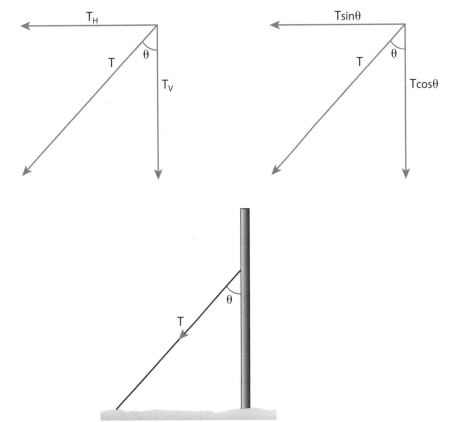

To find the effect of the tension in horizontal and vertical directions, **T** can be split into two components, shown as T_H (the horizontal component) and T_V (the vertical component) in the diagram. You should check that according to the rules of vector addition, $T_H + T_V = T$.

> **KEY POINT**
>
> To find the magnitude of T_H and T_V the following rules apply:
> * the component of a vector **a** at an angle θ to its own angle is acosθ
> * the component of a vector **a** at an angle (90° – θ) to its own angle is asinθ

The application of these rules is shown in the diagram above.

Progress check

1 Work out the resultant of the two forces shown in the diagram.

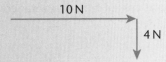

2 A car is being towed with a rope inclined at 20° to the horizontal. The tension in the rope is 350 N.
   Work out the horizontal and vertical components of the tension in the rope.

2 Horizontal component = 329 N   Vertical component = 120 N
1 10.8 N at an angle of 22° to the 10 N force

1.2 Representing and predicting motion

After studying this section you should be able to:

> The term 'curve' means the line drawn on the graph to show the relationship between the quantities. It could be straight or curved.

- *interpret graphs used to represent motion, and understand the physical significance of the gradient and the area between the curve and the time axis*
- *apply the equations that describe motion with uniform acceleration*
- *analyse the motion of a projectile*

The gradients of displacement–time graphs

For a stationary body, displacement is unchanging. The gradient of a displacement–time graph is zero. The velocity is zero.

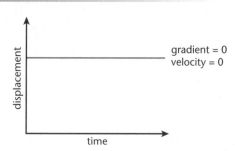

Having a steadily changing displacement is the same thing as having a steady velocity. The displacement–time graph is a straight line. The gradient of the line is constant, and is equal to the velocity.

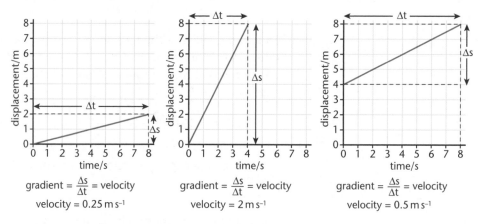

gradient = $\frac{\Delta s}{\Delta t}$ = velocity
velocity = 0.25 m s^{-1}

gradient = $\frac{\Delta s}{\Delta t}$ = velocity
velocity = 2 m s^{-1}

gradient = $\frac{\Delta s}{\Delta t}$ = velocity
velocity = 0.5 m s^{-1}

The motion of an accelerating car is a little more complicated. The displacement changes slowly to start with, and then changes faster and faster. For any point on the graph, the value of the gradient is the same as the value of the instantaneous velocity.

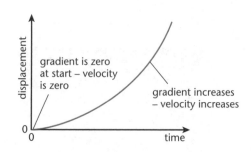

Average and instantaneous values

Calculations of the average velocity over a period of time do not show all of the changes that have taken place. **Instantaneous** values of velocity are defined in terms of **the rate of change** of displacement and are represented by the gradient of the displacement–time graph at a particular instant.

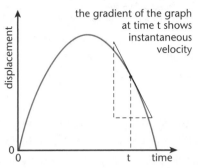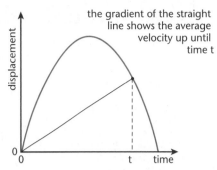

Note that at time t, the instantaneous velocity is negative but the average velocity for the whole journey so far is positive.

> **KEY POINT**
>
> instantaneous velocity = rate of change of displacement
>
> It is represented by the gradient of a displacement–time graph at a particular instant.

The gradient of a displacement–time graph can have either a positive or a negative value, so it shows the direction of motion as well as the speed.

Displacement–time and velocity–time graphs

AQA A	2	Edexcel context	1
AQA B	2	OCR A	1
CCEA	1	OCR B	2
Edexcel concept	1	WJEC	1

It is possible to represent the same motion in different ways. Think of the journey of a stone that is flicked upwards.

Firstly, think in terms of velocity. The initial flick gives it a large upwards velocity but this decreases as the stone rises. For one instant at the top of the journey, the velocity is zero, and then it becomes a downwards velocity. The upwards velocity is positive, and the downwards velocity is negative. The downwards velocity increases in size until the stone has fallen back to where it started.

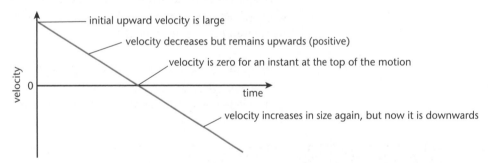

Now think about the displacement. It starts at zero, and is zero again when the stone gets back to where it started. In between, the displacement is always positive. But the displacement does not increase steadily when the stone is first flicked. It increases most rapidly at the start, when the stone is moving fast. Around the top of the journey, displacement changes slowly. The result is a curved line.

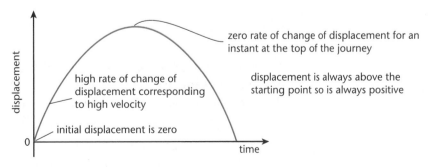

The graphs represent the same motion in different ways. They do not look the same.

Acceleration

In Physics, any change in velocity involves an *acceleration*. This means that an object moving at constant speed, but changing direction, is accelerating.

Any object that is changing its speed or direction is **accelerating**. Speeding up, slowing down and going round a corner at constant speed are all examples of **acceleration**. Acceleration is a measure of how quickly the velocity of an object changes.

> **KEY POINT**
>
> Instantaneous acceleration = rate of change of velocity. It is represented by the gradient of a velocity–time graph at a particular instant.
>
> average acceleration = change in velocity ÷ time $\qquad a = \Delta v \div \Delta t$
>
> Acceleration is a vector quantity and is measured in m s^{-2}.

More about velocity–time graphs

Since acceleration is equal to rate of change of velocity, it is also equal to the gradient of a velocity–time graph.

Remember, the area of a triangl = $\frac{1}{2}$ × base × height.

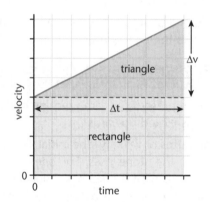

gradient of graph = $\Delta v \div \Delta t$
= acceleration

total displacement can be calculated from area of triangle plus area of rectangle

It is also possible to work out displacement from a velocity–time graph. It is the area 'under the curve'. That is, it is the area between the line and the time, calculated using the units of the two axes.

> **KEY POINT**
>
> The gradient of a velocity–time graph represents the acceleration.
>
> The area between the curve and the time axis of a velocity–time or a speed–time graph represents the distance travelled.

Uniform acceleration

The gradient of the velocity–time graph (see previous section) has a constant value; it represents a constant or **uniform** acceleration. A number of equations link the variable quantities when an object is moving with uniform acceleration. They are:

$v = u + at$
$s = ut + \frac{1}{2}at^2$
$v^2 = u^2 + 2as$
$s = \frac{1}{2}(u + v)t$

The symbols have the meanings already defined:

u = initial velocity
v = final velocity
s = displacement
a = acceleration
t = time

These equations involve a total of five variables but only four appear in each one, so if the values of three are known the other two can be calculated.

When using the equations of uniformly accelerated motion:

- take care with signs: use + and – for vector quantities such as velocity and acceleration that are in opposite directions
- remember that s represents displacement; this is not the same as the distance travelled if the object has changed direction during the motion.

Motion in two dimensions

AQA A	2	Edexcel context	1
AQA B	2	OCR A	1
CCEA	1	OCR B	2
Edexcel concept	1	WJEC	1

There is a simple relationship between the total distances travelled after t, 2t etc. by an object falling vertically. Can you spot it?

People who play sports such as tennis and squash know that, no matter how hard they hit the ball, they cannot make it follow a horizontal path. The motion of a ball through the air is always affected by the Earth's pull.

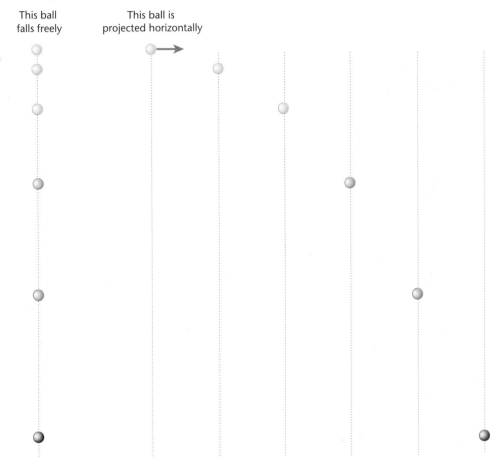

The diagram above shows the results of photographing the motion of a ball projected horizontally alongside one released so that it falls freely. The photographs are taken at equal time intervals.

The free-falling ball travels increasing distances in successive time intervals. This is because the vertical motion is accelerated motion, so the average speed of the ball over each successive time interval increases.

The ball projected horizontally travels equal distances horizontally in successive time intervals, showing that its horizontal motion is at **constant speed**. Vertically, its motion matches that of the free-falling ball, showing that **its vertical motion is not affected by its horizontal motion**.

The motion of any object can be resolved in two directions at right angles to each other. The two motions can then be treated separately.

> **KEY POINT**
>
> When an object has both horizontal and vertical motion, these are independent of each other.

This important result means that the horizontal and vertical motions can be analysed separately:

- for the horizontal motion at constant speed, the equation $v = s \div t$ applies
- for the vertical, accelerated, motion the equations of motion with uniform acceleration apply.

Progress check

1 a The diagram below shows a velocity–time graph. Calculate the displacement and the acceleration for each labelled section of the graph.

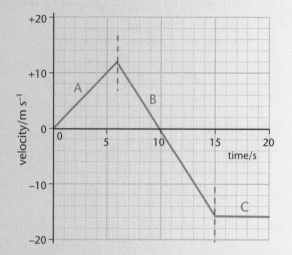

b Is the displacement after 20 s positive or negative? How can you tell this from the graph?

c Sketch a matching displacement–time graph for the whole journey.

2 A clay pigeon is fired upwards with an initial velocity of 23 m s⁻¹. What height does the pigeon reach? Take g = 10 m s⁻².

3 A tennis player is standing 5.5 m from the net. She hits the ball horizontally at a speed of 32 m s⁻¹. How far has the ball dropped when it reaches the net? Take g = 10 m s⁻².

Hint for Q2: what is the velocity of the pigeon at its maximum height?

- Use the constant speed equation to calculate the distance travelled horizontally by the ball in this time.
- Apply the equations of motion with uniform acceleration to the vertical motion to find the time that the ball is in the air. Note that vertically u = 0.

1 a A: 36 m; 2 m s⁻²;
B: 64 m; -3.1 m s⁻²
C: 80 m; 0 m s⁻²
b Negative. The positive area for first 10 s period (sections A and B of the graph) is smaller than the negative area for the second 10 s period.
c Graph
2 26.45 m
3 0.15 m

34

1.3 Force and acceleration

After studying this section you should be able to:

- *predict the effect of resistance to motion on acceleration*
- *distinguish the effects of balanced and unbalanced forces*
- *recall and use the relationship between resultant force, mass and acceleration*
- *distinguish between mass and weight*
- *understand and apply Newton's first and third laws of motion*

LEARNING SUMMARY

Getting going

AQA A	2	Edexcel context	1
AQA B	2	OCR A	1
CCEA	1	OCR B	2
Edexcel concept	1	WJEC	1

> The force that pushes a cycle or other vehicle forwards is called the driving force, or motive force.

Changing motion requires a force. Starting, stopping, getting faster or slower and changing direction all involve forces. One way to start an object moving is to release it from a height and allow the Earth to pull it down. The Earth's pull causes the object to accelerate downwards at the rate of $10 \, \text{m s}^{-2}$.

To set off on a cycle, you push down on the pedal. The chain then transmits this force to the rear wheel. The wheel pushes on the road and, provided that there is enough friction to prevent the wheel from slipping, the push of the road on the wheel accelerates the cycle forwards.

However, as every cyclist knows, this acceleration is not maintained. The cyclist eventually reaches a speed at which the driving force no longer causes the cycle to accelerate. This is due to the **resistive** forces acting on the cycle. The main one is **air resistance**, although other resistive forces act on the bearings and the tyres.

Air resistance or **drag** also affects the motion of an object released and allowed to fall vertically; in fact it opposes the motion of anything that moves through the air. This is due to the air having to be pushed out of the way. The faster the object moves, the greater the volume of air that has to be displaced each second and so the greater the resistive force.

The diagram shows the directions of the forces acting on a cyclist travelling horizontally and a ball falling vertically.

> The single arrow acting on the front of the cyclist represents all the resistive forces acting against the motion.

resistive forces

driving force

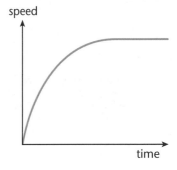

air resistance

weight

The effect of the resistive forces is to reduce the size of the resultant, or unbalanced, force that causes the acceleration. How this affects the speed of the cycle and ball is shown by the graph.

The decreasing gradient of the speed–time graph shows that the acceleration decreases as the resultant force becomes smaller. At a speed where the resistive force is equal to the driving force, the forces are balanced and resultant force is zero, so the object travels at a constant speed.

> When the forces acting on a moving object are balanced, the constant speed is called its terminal speed or terminal velocity.

To summarise:

- an object accelerates if there is a resultant, or unbalanced, force acting on it
- the acceleration is proportional to the resultant force (provided that the mass of the object stays the same) and in the direction of the resultant force.

What else affects the acceleration?

If the mass of an object is doubled, its acceleration is halved for the same pulling force.

The symbol Σ, meaning 'sum of' is used here to emphasise that the relationship applies to the resultant force on an object, and not to individual forces.

As the number of passengers on a bus increases, its ability to accelerate away from the bus stop decreases. Acceleration depends on mass as well as force.

Results of experiments using a constant force to accelerate different masses show that:

the acceleration of an object is inversely proportional to its mass (when force stays the same).

The dependence of acceleration on both the resultant force and the mass is summarised by the relationship:

> **KEY POINT**
>
> resultant force = mass × acceleration
> $$\Sigma F = ma$$
> Where the unit of force, the newton, is defined as the force required to cause a mass of 1 kg to accelerate at $1\,m\,s^{-2}$.

More about balanced and unbalanced forces

It reminds us that space is normal. Where we live is an unusual place in the Universe. Physics has to deal with the normal and also the unusual.

Unbalanced forces always produce acceleration. Balanced forces never produce acceleration.

A car experiences resistive force, in the opposite direction to its motion, but these forces are invisible and all too easy to forget about. The car needs a driving force if it is to keep going at steady speed, so that there is no resultant force. The driving force and the resistive forces are then balanced.

For a spacecraft or any other object in space, where there is no resistance to motion, it is relatively easy to imagine that unbalanced force always produces acceleration and balanced forces never produce acceleration. Exactly the same rules apply here on Earth, we just have to remember to take account of resistive forces.

The force of gravity

In deep space, far from any gravitational pull, g is zero.

In Physics, we use the word weight to mean force of gravity. The weight of a body depends on its location in the Universe. We measure weight in newtons.

We measure mass in kilograms. Mass is not a force. A body has the same mass wherever it goes, as long as it does not lose or gain material.

There is a clash here with everyday language. When people talk about dieting to 'lose weight' then they want there to be less of them. To a physicist, it is mass that they are trying to change.

The problem is that everyday language does not have to distinguish carefully between mass and weight because they are closely related here on Earth. Physics must try to think big, and be universal.

The force of gravity a body experiences depends on its mass, and there is a relatively simple 'conversion factor', which we can write in an equation:

$$g = W/m$$

The conversion factor tells us the force experienced for each kilogram of weight. It is measured in $N\,kg^{-1}$. It is called the **gravitational field strength**. On the Earth's surface, the gravitational field strength is just over $9.8\,N\,kg^{-1}$, which in calculations we often round to $10\,N\,kg^{-1}$.

The acceleration due to gravity, at the Earth's surface, is just over $9.8\,m\,s^{-2}$, which in calculations we often round to $10\,m\,s^{-2}$. The two quantities have identical values. We use g as an abbreviation for acceleration due to gravity as well as for gravitational field strength.

About Newton's laws of motion

AQA A	2	Edexcel context	1
AQA B	2	WJEC	1
CCEA	1		
Edexcel concept	1		

First law

Newton's first law agrees with the statements:
> Unbalanced forces always produce acceleration. Balanced forces never produce acceleration.

The formal statement is:
> An object stays at rest or in uniform motion in a straight line unless there is a resultant (or unbalanced) force acting on it.

Second law

Newton's second law is slightly more technical. It states that resultant force is proportional to rate of change of momentum. Momentum is the mass of a body multiplied by its velocity. A planet has a lot of momentum, a truck accelerating onto a motorway has less, and a drip of water from a tap has less again. All of these bodies are experiencing resultant force, however, and for each one this is proportional to the rate of change of momentum. Using p for momentum, $p = mv$. Newton's second law says:

$$F \propto \Delta(mv)/\Delta t$$

If the mass of a body isn't changing:

$$F \propto m\Delta v/\Delta t$$

The quantity 'change in momentum', $\Delta(mv)$, has its own name – impulse.

Can you see the similarity between this relationship and the equation $F = ma$?

Third law

Newton's third law says that it is impossible for one body (which includes you) to exert a force on another without experiencing a force of the same size in the opposite direction. If you push on a wall, it is not just the wall that experiences force – so do you. The two forces are in opposite directions, and are the same size.

If you push backwards on a floor with your foot, again you experience a force that is the same size as your push. It is how you walk.

A fish flicks water backwards, and experiences a forwards force. A plane blasts air backwards, and experiences a forwards thrust. Newton's third law is involved in all propulsions, whether of animal or machine.

It works on a big scale as well. The force of the Earth on the Moon is the same size as the force of the Moon on the Earth. (The Moon experiences a larger acceleration, because it has less mass.)

The formal statement of Newton's third law is:
> To every action there is an equal and opposite reaction.

Putting it slightly differently:
> If object A exerts a force on object B, then B exerts a force equal in size and opposite in direction on A.

Newton's third law and not Newton's third law

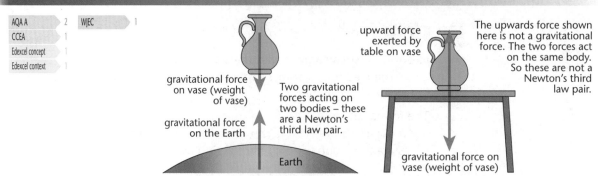

AQA A	2	WJEC	1
CCEA	1		
Edexcel concept	1		
Edexcel context	1		

gravitational force on vase (weight of vase)

gravitational force on the Earth

Two gravitational forces acting on two bodies – these are a Newton's third law pair.

Earth

upward force exerted by table on vase

The upwards force shown here is not a gravitational force. The two forces act on the same body. So these are not a Newton's third law pair.

gravitational force on vase (weight of vase)

Newton's third law

The Earth exerts a gravitational force on a vase. The vase exerts a gravitational force on the Earth. The vase accelerates (much) more than the Earth because of its (much) smaller mass.

- The two forces are: of the same type – both forces of gravity.
- The two forces act on different objects.
- The two forces are inevitable consequences of each other – there is no escape from Newton's third law.

NOT Newton's third law

A table under the vase exerts an upwards force on it. But these are not a 'Newton's third law pair' of forces. The gravitational pair are still there, and the new upwards force doesn't affect them.

- Now, the two forces acting on the vase are not both of the same type. The upwards force due to the table is not a gravitational force. This is enough to show that they are not a 'Newton's third law pair'.
- The weight of the vase and the upwards force of the table act on the same body, the vase. Again, these forces can't be Newton's third law pairs.
- The upwards force of the table is not an absolutely inevitable consequence of the force of gravity pulling the vase down.

Progress check

1 a The total mass of a cyclist and her cycle is 85 kg.
   Calculate the acceleration as she sets off from rest if the driving force is 130 N.

   b The driving force remains constant at 130 N. What is the size of the resistive force when she is travelling at terminal velocity?

2 A fully laden aircraft weighs 300 000 kg. It accelerates from rest to a take-off speed of 70 m s⁻¹ in 20 s.

   a Calculate the acceleration of the aircraft.

   b Calculate the size of the resultant force needed to cause this acceleration.

   c Explain why the force from the engines must be greater than the answer to **b**.

3 A tightrope walker stands in the middle of a rope.

   a What exerts the upward force on the tightrope walker?

   b Explain why it is not possible for the rope to be perfectly horizontal.

4 Explain how you can tell from a vector diagram that an object is in a state of rest or uniform motion.

4 The arrows that represent the vectors form a closed figure; i.e. the last arrow finishes where the first one starts.

   b A rope can only pull in its own direction. For the tension to have a vertical component, the rope must be inclined to the horizontal.

3 a The tension in the rope.

   c The force also has to act against air resistance and friction.

2 a 3.5 m s⁻²   b 1.05 × 10⁶ N

1 a 1.53 m s⁻²   b 130 N

1.4 Energy and work

After studying this section you should be able to:

- *identify situations where a force is working*
- *understand that energy can transfer from system to system and recognise that most energy transfers involve some dissipation of energy*
- *explain that some systems act as energy stores*
- *explain that during some energy transfers, physical work can be done, and in real energy transfers heating takes place*
- *use Sankey diagrams to illustrate energy transfer processes*
- *apply the principle of conservation of energy*
- *predict rate of transfer of energy by heating (thermal transfers)*
- *classify energy resources*
- *explain global energy balance and equilibrium temperature, and the greenhouse effect*

Working

AQA A	2	Edexcel context	1
AQA B	2	OCR A	1
CCEA	1	OCR B	2
Edexcel concept	1	WJEC	1

Every event requires **work** to make it happen. Any force that causes movement is doing work. Pushing a supermarket trolley is working, as is throwing or kicking a ball. However, holding some weights above your head is not working; it may cause your arms to ache, but the force on the weights is not causing any movement!

How much work a force does depends on:
- the size of the force
- the direction of movement
- the distance that an object moves.

The phrase, 'distance moved in the direction of the force', is used here instead of 'displacement' to emphasise that the force and displacement it causes are measured in the same direction.

KEY POINT

The work done by a force that causes movement is defined as:

work = average force × distance moved in the direction of the force
$$W = F \times s$$

Work is measured in joules (J) where 1 joule is the work done when a force of 1 N moves its point of application 1 m in the direction of the force.

Here are some examples of forces that are working and forces that are not working.

A pylon that supports electricity transmission cables is not working, but a wind that causes the cables to move is!

The shelf is not moving, so no work is being done.

The tension in the string is not working as there is no movement **in the direction of the force.**

This force causes movement **in the direction of the force.**

The horizontal component of the force on the log is doing the work here.

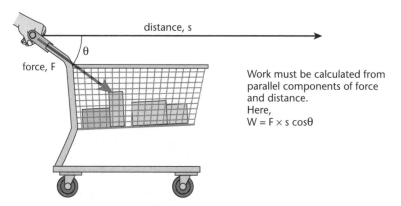

distance, s

force, F

θ

Work must be calculated from parallel components of force and distance.
Here,
$W = F \times s \cos\theta$

In circular motion, the direction of the velocity is along a tangent to the circle.

In the case of the stone being whirled in a horizontal circle, the stone's velocity is always at **right angles** to the force that maintains the motion, so the force is not working.

Although the force pulling the log and the movement it causes are not in the same direction, the force on the log has a component in the same direction as the log's movement. In calculating the work done the horizontal component of the force is multiplied by the horizontal distance moved by the log.

Energy stores and transfers

AQA A	2	Edexcel context	1
AQA B	2	OCR A	1
CCEA	1	OCR B	2
Edexcel concept	1	WJEC	1

'Systems' store energy. A reservoir full of water in the mountains acts as en energy store, as does a hammer raised up above a nail. A flying bullet stores energy, as does a flywheel. A fuel, together with an oxygen supply, acts as an energy store. So does an electrical battery. A cylinder full of hot gas is an energy store.

In all cases, energy can transfer out from the store into other systems. Sometimes the energy transfers to another storage system, and sometimes it is dissipated in the surroundings. In that case, the surroundings are heated.

Energy transfers always involve working or heating or both of these. Working and heating are both energy transfer processes. They can be called mechanical and thermal processes.

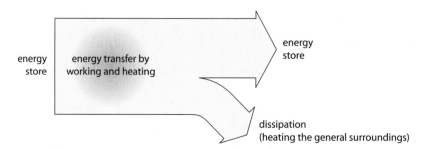

energy store

energy transfer by working and heating

energy store

dissipation
(heating the general surroundings)

A representation of energy flow like this is sometimes called a Sankey diagram.

The following are some examples of everyday energy transfers.

A bus accelerates away from a bus stop and then maintains a steady speed

As the bus speeds up, energy stored in the fuel–oxygen system transfers to energy stored by the motion of the bus. This store is called the kinetic energy of the bus. Other energy from the fuel–oxygen transfers to the surroundings, heating them up just a little. Some energy is transferred by the hot exhaust gases, and some is transferred by the action of forces of resistance. This energy becomes thinly spread, and we say it has dissipated.

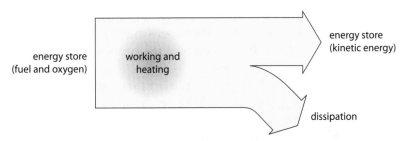

Once the bus is moving at steady speed, there is no more increase in its kinetic energy, so all of the energy transferred from the fuel and oxygen is being dissipated in the surroundings.

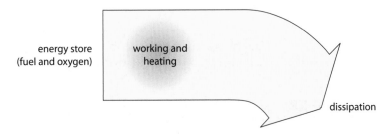

An electric motor lifts a load at a steady speed

An electric motor is an energy transfer device.

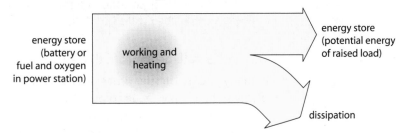

As the motor lifts a load, the load gains gravitational potential energy, which is one way in which a mechanical system can store energy. Its gravitational potential energy changes because its height above the Earth's surface changes. Except right at the start and finish, there is no gain or loss of kinetic energy of the rising load, since its speed is constant. There is, though, some dissipation due to resistive forces (which include frictional forces).

A filament lamp lights a room

When a filament lamp is switched on it takes a fraction of a second to reach its operating temperature.

Once the lamp has reached its steady state, of the 60 J of energy passing into the lamp each second from the electricity supply, typically 3 J passes out as light. This is shown in the diagram. The Sankey diagram gives a visual indication of the relative proportions of energy transferred into a desirable output and wasted.

In the steady state, the parts of the lamp are emitting energy and absorbing energy at the same rate, so there is no change in temperature.

Sankey diagrams are a useful way of showing the energy flow through a process where there are several stages, for example the energy flow through a power station.

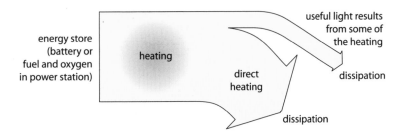

The 'wasted' energy is transferred to the surroundings in two main ways:
- from the glass envelope to the surrounding air by conduction and convection
- from all parts of the lamp as non-visible electromagnetic radiation (mainly infra-red).

The light and the infra-red radiation both cause heating when they are absorbed by other objects, so all the energy input to the lamp ends up as heat!

Work and energy

AQA A	2	Edexcel context	1
AQA B	2	OCR A	1
CCEA	1	OCR B	2
Edexcel concept	1	WJEC	1

In a purely mechanical system where there is no energy transfer by heating, the energy transfer from a store or source of energy is equal to the amount of work done.

> **KEY POINT**
>
> Energy is sometimes described as the ability, in an ideal mechanical system, to do work. Like work, energy is measured in joules, J.

Conservation of energy

AQA A	2	Edexcel context	1
AQA B	2	OCR A	1
CCEA	1	OCR B	2
Edexcel concept	1	WJEC	1

Note that the total width of the arrows in the Sankey diagrams is always the same before and after the transfer, to show that the total amount of energy stays the same. Like mass and charge, energy is a conserved quantity.

> **KEY POINT**
>
> The principle of conservation of energy states that:
> energy cannot be created or destroyed.

 Take care not to confuse 'conservation of energy' with 'conservation of energy resources'. Energy resources such as coal can be used up, and conservation in this context means preserving them as long as possible.

This simple statement means that energy is never used up.

In many energy transfer processes the amount of energy at each stage cannot easily be quantified; it is difficult to put a figure on the total kinetic and potential energy of the moving parts in an engine for example. There are simple formulae for calculating the kinetic energy and gravitational potential energy of individual objects.

 The formula for change in gravitational potential energy is only valid for changes in height close to the Earth's surface, where g has a constant value.

> **KEY POINT**
>
> $$\text{kinetic energy} = \tfrac{1}{2} \times \text{mass} \times (\text{speed})^2$$
> $$E_k = \tfrac{1}{2} mv^2$$
> $$\text{change in gravitational potential energy} = \text{weight} \times \text{change in height}$$
> $$\Delta E_p = mg\Delta h$$

The principle of conservation of energy applies to all energy transfers.

When a streamlined object is falling freely in the atmosphere at a low speed, the resistive forces are small and so little energy is transferred to the surroundings. In this case the principle of conservation of energy can be applied to the transfer of gravitational potential energy to kinetic energy. When a parachutist is falling at a constant speed conservation of energy still applies, but the energy transfer is from gravitational potential energy to heat in the atmosphere.

Energy transfer of falling objects

a ball falling freely
towards the Earth

a parachutist falling
at constant speed

gravitational potential
energy → kinetic energy

gravitational potential
energy → heat in the
surrounding air

Potential energy, kinetic energy, and predicting the speed of a falling body

AQA A	2	Edexcel context	1
AQA B	2	OCR A	1
CCEA	1	OCR B	2
Edexcel concept	1	WJEC	1

For a body that falls freely, without resistance to motion:

potential energy lost = kinetic energy gained

After falling height Δh and acquiring velocity v:

$$mg\Delta h = \tfrac{1}{2}mv^2$$
$$2g\Delta h = v^2$$
$$v = \sqrt{2g\Delta h}$$

This equation provides a way of predicting velocity of fall from height fallen.

Power

AQA A	2	Edexcel context	1
AQA B	2	OCR A	1
CCEA	1	OCR B	2
Edexcel concept	1	WJEC	1

Typically, a car engine has an output power of 70 kW, compared to 1.0 kW for a hairdryer and 2 W for a clock.

Power is a measure of the work done or energy transferred each second.

> **KEY POINT**
> Power is the rate of working or energy transfer.
> Average power can be calculated as work done ÷ time taken, $P = \Delta W \div \Delta t$
> Power is measured in watts (W), where 1 watt is a rate of working of 1 joule per second ($J\,s^{-1}$).

Power and efficiency

AQA A	2	Edexcel context	2
AQA B	2	OCR A	1
CCEA	1	WJEC	1
Edexcel concept	2		

Do not confuse power with speed. A double-decker bus may be slower than many cars, but it can transport 60 passengers a given distance much faster than any car!

In choosing or designing a machine to do a particular job, important factors to consider include the **power input** and the **power output**. You would expect a hairdryer with a high power output to dry your hair faster than one with a low power output, but one with a high power input may not do the job any faster than one with a lower power input, it may just be less efficient!

Steam trains have a very high power input, but a very low efficiency at transferring this to power output

A comparison of the power output and power input of a device tells us the efficiency. A ratio is a mathematical comparison, so we make a ratio of useful power output and total power input, and we call that the efficiency of the device.

That is:

efficiency = useful power output ÷ total power input

It is quite common to multiply the answer here by 100 to give a percentage efficiency.

A fluorescent lamp, for example, can be described as having an efficiency of 20%.

Energy and efficiency

AQA A	2	Edexcel context	2
AQA B	2	OCR A	1
CCEA	1	WJEC	1
Edexcel concept	2		

For a particular period of time, we can measure the useful energy output and the total energy input of a device, and perform the division to find the efficiency.

efficiency = useful energy output in a specified time ÷ total energy output in the same time

So, for the filament lamp that has an energy input of 60 J in 1 second, and a useful energy output of 3 J, also in 1 second:

efficiency = 3 ÷ 60 = 0.05

We can multiply this by 100, and describe the lamp as being 5% efficient.

Thermal energy transfers

AQA B ▶ 2

Newton's law of cooling

If two bodies have different temperatures then energy normally flows from the one at the higher temperature to the one at the lower temperature. So a body that has a higher temperature than its surroundings cools down. If the surroundings are significantly larger than the body, the energy will spread out and the temperature rise of the surroundings will be small.

Newton's law of cooling states that the rate of change of temperature of a cooling body is proportional to its temperature difference with the surroundings.

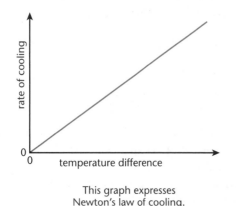

This graph expresses Newton's law of cooling.

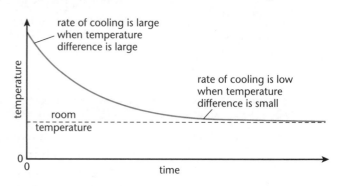

This graph, for a heated body cooling to room temperature, is a consequence of Newton's law of cooling.

Factors affecting rate of energy transfer through a material

Suppose that a solid material separates two bodies that are at different temperatures. There will be a tendency for energy to flow through the material, from the hotter to the cooler body.

Factors that affect the rate of energy transfer are:

- the **thermal conductivity** of the material, which is a measure of its ability to transfer energy by thermal conduction, k
- the area through which the energy can flow, A
- the temperature gradient across the material, which is the total temperature difference divided by the thickness of the material, $\Delta T / \Delta x$

Rate of energy transfer by thermal conduction $= k\,A\,\Delta T / \Delta x$

The quantity $k/\Delta x$ is called the **U value** of the insulation.

Energy resources

AQA B ▶ 2

The Sun is almost, but not quite, the only source of energy for life on Earth. Energy from the Sun not only keeps us alive, but drives the temperature differences that give rise to ocean currents and winds. Energy from the Sun evaporates water, some of which returns as rain or snow to the land to create rivers. Dammed rivers provide hydroelectric generation systems, or HEP.

Wind turbines:

Maximum power of turbine motion from wind in ideal conditions

$$= \tfrac{1}{2}\pi r^2 \rho v^3$$

where r is the radius of the turbine, ρ is the density of the air and v is the air speed.

There are some sources of energy other than the Sun, however. The most used of these are nuclear resources, based on fission of large nuclei that were created in previous generations of stars. The mass of these nuclei, a small proportion of the total that they have, ceases to exist during the fission process, and the mass loss is matched by energy release.

Geothermal energy resources take advantage of the heat of the inner Earth that is due to the natural radioactivity of our planet.

Tides slowly drain energy away from the relative motions of Earth and Moon, and we can use the flowing water to generate electricity.

Our distance from the Sun is important. The power of sunlight decreases rapidly with distance from the Sun, obeying the **inverse square law**.

Supposing the Sun acts as a single point, the intensity of energy received at an area at distance r from the Sun, measured as the energy arriving per unit time per unit area, is related to the Sun's total power output (P) by the equation:

Intensity, $I = P / 4\pi r^2$

Planetary energy balance

AQA B   2

No body is totally isolated from any other, and no body is so cold that it could never become colder. All bodies emit some thermal radiation, and they also receive it from other bodies.

For a body to have a constant temperature it must emit and receive energy at the same rate. A body reaches a natural **equilibrium temperature**.

If it were to warm up, it would emit energy faster, and this opposes the warming.

If it cooled, the rate of emission would be less, so again there would be a tendency for balance or equilibrium to be restored.

Thus the planet Mercury has a natural equilibrium average surface temperature, as does the planet Venus.

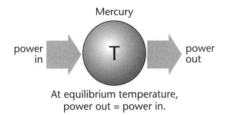

Mercury

power in → T → power out

At equilibrium temperature,
power out = power in.

Venus, however, has an atmosphere that resists the outward flow of radiant energy more than it resists the inward flow. Equilibrium is reached when outward flow and inward flow are the same and the planet's temperature has risen so that outward flow can be large enough. This what is called the greenhouse effect.

Venus

power in → T → power out

If energy escapes more slowly than it arrives, then temperature rises.

Venus

power in → T → power out

A hotter object tends to emit energy more rapidly. Temperature rises until power out = power in. This produces a new (hotter) equilibrium temperature.

There is a natural greenhouse effect on Earth, as well. The problem is that human activity has increased the concentration of greenhouse gases in the atmosphere – particularly carbon dioxide. It now seems very likely that human activity is producing an additional greenhouse effect, which is in turn very likely to produce climate changes that are hard to predict and that will escalate for as long as humans add to the greenhouse gases.

Progress check

1 A 240 N force is used to drag a 50 kg mass a distance of 5.0 m up a slope (see diagram). The mass moves through a vertical height of 1.8 m. Take $g = 10\,\text{m s}^{-2}$. Calculate:

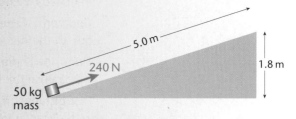

   a The work done on the mass.
   b The gravitational potential energy gained by the mass.
   c The efficiency of this way of lifting the mass through a vertical height of 1.8 m.
   d The time taken to drag the mass up the slope by a motor that has an output power of 150 W.
   e Draw a Sankey diagram for the process.

2 A car pulls a trailer with a force of 750 N, transmitted through the tow bar. The tow bar is inclined at 20° to the horizontal. Calculate the work done by the car in pulling the trailer through a horizontal distance of 200 m.

3 a Explain why the planet Venus has a constant average temperature despite continuous arrival of energy from the Sun.
   b Explain why the temperature is much higher than would be expected for a similar planet in similar orbit but with no atmosphere.

1 a 1200 J
   b 900 J
   c 0.75
   d 8.0 s
   e Diagram should show relative amounts of gravitational potential energy and energy dissipated.
2 1.4×10^5 J
3 a Planet radiates out to space at same rate as energy is received.
   b There is a greenhouse effect / gases resist outward flow of energy / temperature must be high in order that rate of emission is high enough.

1.5 Vehicles in motion

After studying this section you should be able to:

- *discuss the forces involved in braking and towing*
- *explain the principles of seat belts, air bags and crumple zones*
- *interpret and explain data relating to vehicle stopping distances*
- *do calculations involving motive power*
- *explain the difference between laminar and turbulent flow*
- *calculate the drag force acting on a vehicle and explain how it varies with speed*
- *explain why objects reach a terminal velocity*
- *describe how balance of forces, horizontally and vertically, is achieved in flight*

LEARNING SUMMARY

Forces and braking

OCR A   1

Vehicles that travel on wheels rely on friction between the tyres and the road surface to provide the driving force. Ice on a road surface and wet leaves on railway lines can reduce the friction force, resulting in the wheels spinning round as they slide over the surface.

Friction is also used in braking systems to bring cycles and motor vehicles to a halt. Two common types of brake used are **drum brakes** and **disc brakes**. These are shown in the diagram.

> Most modern vehicles use disc brakes because of their greater efficiency.

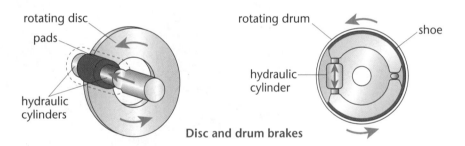

Disc and drum brakes

> A deceleration is a negative acceleration. In this case the acceleration is in the opposite direction to the vehicle's velocity, so it has the opposite sign.

When the driver applies the brakes, high friction material in the shoe or pad is pressed against the drum or disc. The friction force exerts a torque in the opposite direction to the torque that drives the wheel. During braking the tyre pushes **forwards** on the road surface, causing the road surface to push **backwards** on the tyre. The resultant force in the backwards direction causes the vehicle to decelerate.

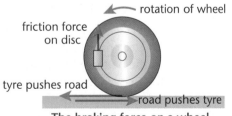

The braking force on a wheel

Braking forces, like driving forces, rely on friction. The size of the friction force depends on the nature of both the road and the tyre surface. The greater the friction force, the shorter the distance required to stop the vehicle.

Grand Prix drivers match their tyres to the road conditions. Excessive friction wastes fuel and reduces the acceleration and top speed. However, a car fitted with tyres suitable for dry road conditions can be out of control if there is a sudden shower of rain.

Tyre **tread** is designed to prevent a layer of water building up between the tyre and the road. The presence of water reduces the friction force, reducing the effectiveness of steering, drive and braking. If there is insufficient tread on a tyre, it cannot push the water out of the way quickly enough, as the channels in the tyres are too small for all the water to pass through.

Car safety

OCR A 1

When the velocity of a car changes, that of the people inside it has to change too. During normal driving, the force to accelerate the driver and passengers comes from the seat; it pushes the person forwards to cause an increase in speed and the friction force between the person and the seat is sufficient to slow the person down or cause a change in direction.

During rapid deceleration, a passenger appears to be 'thrown' in the opposite direction to the change in velocity.

During hard braking or in a collision, friction is not enough to match the car's deceleration. Viewed from the outside of the car, a passenger appears to have been thrown forwards. Despite what the television advertisements say, this is **not** the case! The photograph shows what can happen to a passenger in a collision. **Seat belts** prevent this from happening. However, a rigid seat belt that causes a passenger to decelerate at the same rate as the car could prove fatal in a head-on collision.

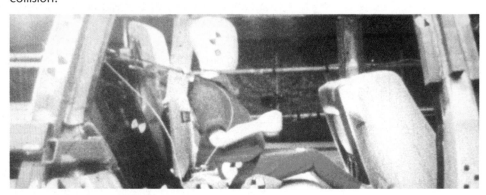

To be effective a seat belt should:
- restrain the passenger and prevent collisions with the inside of the car
- allow the passenger to come to rest over a longer time period than that taken by the car.

Excessive force from a seat belt can break the ribs and damage internal organs.

To achieve these conflicting aims, seat belts are designed so that they stretch sufficiently to allow the passenger to carry on moving for a short time after the car has stopped, but not so much that would result in the passenger hitting the windscreen or the seat in front.

Seat belts can still inflict severe injuries during a collision. The amount of space to allow stretching, particularly in front of a driver, is limited so the restraining force from the belt can be large. There is also a design conflict in deciding on the width of the belt; wide belts exert less pressure than narrow ones but are less comfortable to wear.

There is less space in front of the driver than there is in front of the passengers because of the steering wheel. In modern cars this is designed to collapse on impact.

Airbags are misnamed as they do not use air. An inert gas such as nitrogen or argon is used to inflate the bag within 0.02 s of a collision.

Airbags are designed to restrain a driver and passengers without any risk of causing physical damage to the person. The bag only operates during a rapid deceleration such as a head-on collision. This releases a gas which causes the bag to inflate and surround the driver or passenger like a cushion. The force exerted on the person is similar to that from a seat belt, but because it is over a much larger area the pressure is smaller.

Modern cars and railway carriages are also designed so that parts of them crush on impact. Although the part of the vehicle containing passengers is rigid to give them protection, the front and rear are **crumple zones**.

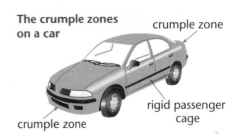

The crumple zones on a car

crumple zone

rigid passenger cage

crumple zone

As the name implies, the crumple zone should squash during a collision. This has two effects:

- it increases the time it takes to stop the vehicle, reducing the deceleration and the force
- it absorbs much of the kinetic energy of the moving vehicle, so that it does not bounce backwards, causing further injury to passengers.

Stopping distances

OCR A ▶ 1

In the highway code there is a chart that shows the **stopping distances** in good conditions for a car travelling at different speeds. There are two components to the stopping distance at a particular speed, **thinking distance** and **braking distance**.

Thinking distance is how far the vehicle travels while the driver is reacting. The two factors that determine this are:

Drugs, including alcohol, nagging children and a loud hi-fi can all increase a driver's reaction time.

- reaction time
- vehicle speed.

A driver's reaction time depends on a number of things, including the state of the driver and the nature of the hazard, but a driver should be able to react to an unexpected hazard within 0.6 s.

As the diagram below shows, this means that doubling the vehicle speed doubles the thinking distance.

During the driver's reaction time the vehicle continues at a constant speed, so **distance travelled = speed × time**. Assuming a constant reaction time, this means that:

thinking distance is proportional to speed.

This is not the case with braking distance. As with reaction time, a number of factors could affect this, including the nature and condition of the road surface, the tyres and the brakes. The following calculations to work out the braking distances from speeds of $10\,\text{m s}^{-1}$ and $20\,\text{m s}^{-1}$ assume a deceleration of $8.3\,\text{m s}^{-2}$.

initial speed/m s⁻¹	10	20
time taken to brake/s	1.2	2.4
average speed during braking/m s⁻¹	5	10
braking distance/m = average speed × time	6	24

This shows that doubling the speed quadruples the braking distance. Not only is the average speed during braking doubled, so is the time it takes to stop, so:

braking distance is proportional to (speed)².

The diagram shows how stopping distance is related to speed:

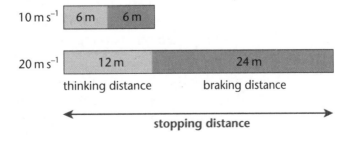

The overall effect of speed on stopping distance is approximately 'doubling the speed trebles the stopping distance'.

Vehicles and motive power

OCR B   2

The expression (F × s) ÷ t, representing **work done** ÷ **time taken** is also equal to **force** × **velocity**. This gives an alternative way of calculating power in situations where the force and speed in the direction of the force (the component of the velocity in that direction) are known:

> Power = force × velocity
>
> $P = F v$

KEY POINT

This expression can be applied to the motive power of a vehicle, where F is the driving force and v is the speed of the vehicle.

For example, if a force of 600 N is used to tow a trailer at a speed of 12 m s⁻¹, then the motive power required is 600 N × 12 m s⁻¹ = 7200 W.

Streamlines and turbulence

AQA B   2
Edexcel context   1
OCR A   1

In a wind tunnel the air moves over a stationary object such as a car body to model the movement of the car through the air. This is a valid model, as it is the relative speed of the air and the object that determines the pattern of flow.

Viscosity is a measure of the resistance of a fluid to flowing. Syrup and tar are viscous fluids; hydrogen has a very low viscosity.

To be as fuel-efficient as possible, a car, train or aircraft needs to be designed to minimise the resistive forces acting on it. To investigate the resistive or **drag** forces due to movement through the air, a car body is placed in a **wind tunnel**. Here the car body is held stationary and the air moves around it; vapour trails show the air flow over the car body. The diagram below shows the air flow at a low speed over a car body. The flow is said to be **laminar** or **streamlined**.

In laminar flow:

* the air moves in layers, with the layer of air next to the car body being stationary and the velocity of the layers increasing away from the car body
* particles passing the same point do so at the same velocity, so the flow is regular
* the drag force is caused by the resistance of the air to layers sliding past each other
* more **viscous** air has a greater resistance to relative motion and exerts a bigger drag force
* the drag force is proportional to the speed of the car relative to the air.

The air flow over a car travelling at low speed

Turbulence occurs at higher speeds

As the speed of the air passing over the car body is increased, the flow pattern changes from laminar to **turbulent**. This is shown in the diagram above.

The changeover from laminar flow to turbulent occurs at a speed known as the **critical velocity**. Turbulent flow causes much more drag than laminar flow.

In turbulent flow:

* the air flow is disordered and irregular
* the drag force depends on the **density** of the air and not the viscosity
* the drag force is proportional to the (speed)² where speed is relative to the air.

A fluid exerts a **viscous drag** on a body that is moving through it. The extent to which a fluid does this is quantified in terms of its viscosity. A more viscous fluid, like heavy oil or treacle, exerts a larger viscous drag than does a less viscous one, such as air. The size of the viscous drag is related to fluid viscosity, N, body size represented as effective sphere radius, r, and body velocity, v, by an equation that is known as **Stokes' Law**:

$F = 6\pi Nrv$

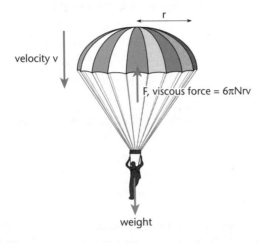

Terminal velocity

If an object is falling through a fluid, e.g. air or water, as its speed increases, the drag on it will also increase. Eventually a speed is reached where the upward force will equal the weight of the object. As there is no net force on the object the acceleration will be zero. The object will fall at a constant velocity. This velocity is called the **terminal velocity**.

Similarly for a car, or other vehicle, when it reaches a speed where the sum of all the forward forces, i.e. the thrust, equals the sum of all the resistive forces, the drag, the vehicle will move at a terminal velocity.

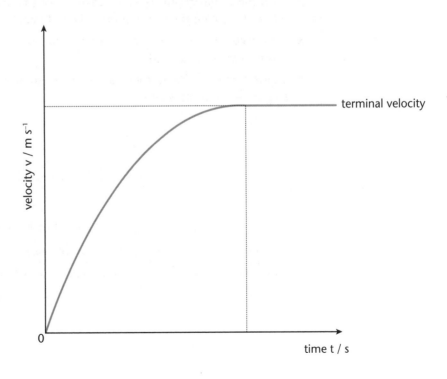

Forces in flight

AQA B 2
OCR A 1

Drag is the force of air resistance that opposes motion. For motion at steady speed, drag must be balanced by a driving force or a thrust provided by an aircraft engine.

Lift is the upwards force on the wings of an aircraft, which starts with the flow of air across the wing surfaces. The wing shape is called an aerofoil. Its flow creates a lower pressure on the upper wing surface than on the lower. For steady flight, lift must be big enough to balance weight.

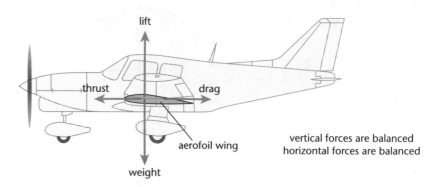

It is easier to consider the vertical forces and the horizontal forces independently, rather than as a complete set of four. Knowledge of whether the forces are balanced or unbalanced allows the flight of the aircraft to be predicted.

Progress check

1 A car is travelling at the legal maximum speed of 28 m s^{-1} on a single carriageway. The driver notices a hazard in the road ahead.

   a If the driver's reaction time is 0.5 s and the car decelerates at 7 m s^{-2} during braking, calculate the stopping distance.

   b Calculate the size of the force required to decelerate the car if its total mass is 850 kg.

2 A car has a maximum speed of 65 m s^{-1}. At this speed, the motive power is 90 kW. Calculate the value of the driving force.

3 State **two** differences between laminar flow and turbulent flow of air over an object such as a car body.

4 Suggest why it is dangerous for an aircraft to attempt to take off when there is a layer of ice on its wings.

4 the ice increases the turbulence, reducing the lift.

viscosity of the air.
proportional to the relative speed of the object and the air; in laminar flow the drag is proportional to the
drag force is due to resistance of the layers of air sliding over each other; in laminar flow the drag force is
3 Any two from: laminar flow is regular; in laminar flow layers of air slide over each other; in laminar flow the

2 1385 N

1 a 70 m   b 5950 N

1.6 More effects of forces

After studying this section you should be able to:

- *calculate the moment of a force and apply the principle of moments to a stable object*
- *calculate the pressure due to a force*
- *calculate the density of a solid, liquid or gas*

LEARNING SUMMARY

The turning effect

AQA A	2
CCEA	1
OCR A	1
WJEC	1

When forces are used to open doors, steer and pedal bicycles or turn on taps, they are causing turning. The effect that a force has in turning an object round depends on:

- the size of the force
- the perpendicular (shortest) distance between the force line and the pivot (the axis of rotation).

Both of these factors are taken into account when measuring the turning effect, or moment, of a force.

> The 'line of action' of a force is the line drawn along the direction in which the force acts.

> **KEY POINT**
> Moment of a force = force × perpendicular distance from line of action of force to pivot.
> The moment of a force, also known as torque, is measured in N m.

The diagram shows how the same force used to open a door can produce very different effects, according to the direction in which it is applied.

> The moment of a force is a vector that can only have one of two directions; either clockwise or anticlockwise.

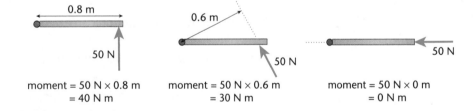

moment = 50 N × 0.8 m = 40 N m

moment = 50 N × 0.6 m = 30 N m

moment = 50 N × 0 m = 0 N m

A question of balance

AQA A	2
CCEA	1
OCR A	1
WJEC	1

When turning on a tap or steering a bicycle, two forces are normally used. The forces act in opposite directions, but they each produce a moment in the same direction. A pair of forces acting like this is called a couple. The combined moment is equal to the sum of the moments of the individual forces.

If the forces that make the couple are equal in size, the moment of the couple = size of one force × shortest (perpendicular) distance between the force lines.

Forces in structures

AQA A	2
CCEA	1
OCR A	1
WJEC	1

If a beam or rod is uniform its mass is evenly distributed so that it is the same as if all the mass was concentrated at a single point at the centre of the beam. The weight is the force acting vertically downwards from this point.

The resultant force and resultant moment at any point in a stable structure are both zero. Either of these principles can be used to find the values of forces F_1 and F_2 in the ropes that support the beam shown in the diagram overleaf.

The resultant force and resultant moment both have to be zero for the structure to be stable.

Resolving horizontal and vertical components will always give the same answers as taking moments.

- Both F_1 and F_2 could be found by resolving the vertical components of all the forces (which must balance) and the horizontal components of all the forces (which must also balance).
- Since F_1 does not have a moment about A, taking moments about this point gives a relationship between F_2 and the forces we know.
- Similarly, taking moments about B gives a relationship between F_1 and the known forces.

To take moments about A, F_2 needs to be resolved into horizontal and vertical components. The moment of the horizontal component is zero and that of the vertical component is in the anticlockwise direction. The equation is:

$$F_2 \cos 60° \times 4.0\,m = 200\,N \times 2.0\,m + 80\,N \times 1.0\,m$$
$$F_2 = 480\,N\,m \div (\cos 60° \times 4.0\,m) = 240\,N.$$

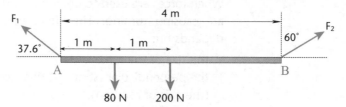

Pressure

OCR A   1

When a force is applied to a solid object such as a drawing pin or a nail, the force is transmitted through the solid. Factors that determine the effect of the force in cutting or piercing include:

- the size of the force
- the area of contact.

If the force that causes the pressure is not acting in a direction normal to the surface, then the normal component of the force is used to calculate the pressure.

These two factors together are used to calculate the pressure.

> Pressure = normal force per unit area
> $$p = F/A$$
> Pressure is measured in pascal (Pa) where $1\,Pa = 1\,N\,m^{-2}$
>
> **KEY POINT**

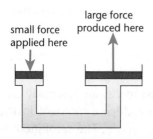

small force applied here

large force produced here

Whereas a solid transmits a force in the direction in which it is applied, fluids (liquids and gases) transmit pressure. The pressure at any point in a fluid has the same value in all directions. The transmission of pressure by a fluid is used in hydraulic machinery, where the force exerted by the fluid is determined by its pressure and the area over which it acts. This enables a large force to be exerted by the application of a small force that causes the pressure in the fluid. The principle of a hydraulic lifting device is shown in the diagram opposite.

Density

AQA A	2	Edexcel context	1
AQA B	2	OCR A	1
CCEA	1	OCR B	1
Edexcel concept	1	WJEC	1

The density of a material is a measure of how close-packed the particles are. Gases at atmospheric pressure are much less dense than solids and liquids because the particles are more widespread.

Typical densities of some everyday materials in $kg\,m^{-3}$ are:

air 1.2
water 1000
concrete 5000
iron 9000

> Density of a material = mass per unit volume.
> $$\rho = m/V$$
> Density is measured in $kg\,m^{-3}$ or $g\,cm^{-3}$.
>
> **KEY POINT**

Density is measured by dividing the mass of a specimen of the material by its volume.

Forces in fluids

A ship or an iceberg experiences an upwards force exerted, in effect, by the water. This is called **upthrust**. Any fluid exerts an upthrust on a body immersed (or partly immersed) in it. The size of the upthrust is equal to the weight (in newtons, of course) of the fluid that the body displaces. This statement is called **Archimedes' principle**.

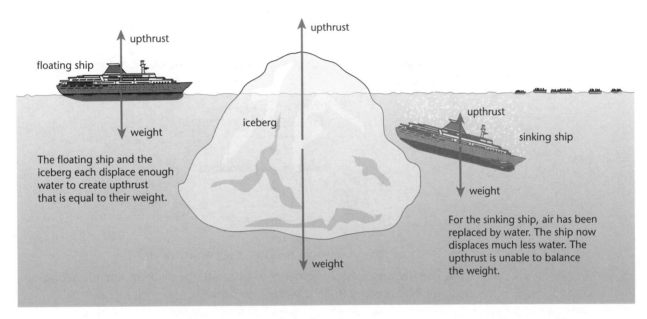

floating ship — upthrust / weight

iceberg — upthrust / weight

sinking ship — upthrust / weight

The floating ship and the iceberg each displace enough water to create upthrust that is equal to their weight.

For the sinking ship, air has been replaced by water. The ship now displaces much less water. The upthrust is unable to balance the weight.

Note that when floating ice melts, its mass and weight do not change, so it takes up exactly the same volume as the volume of water it displaced when solid. Melting sea ice, therefore, has no effect on global sea levels.

Note also that the Archimedes' principle applies to bodies in gases as well as in liquids. For a body to float in air, it must displace a volume of air with a greater weight than its own. That is, it must be less dense than the air.

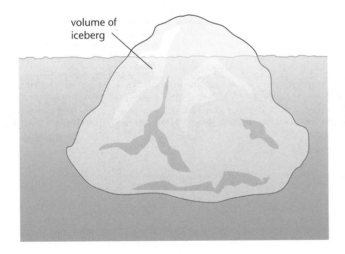

volume of iceberg

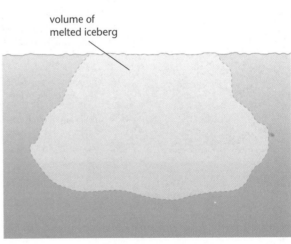

volume of melted iceberg

Progress check

1. Use the principle of moments to calculate the tension in the rope, T, in the diagram below.

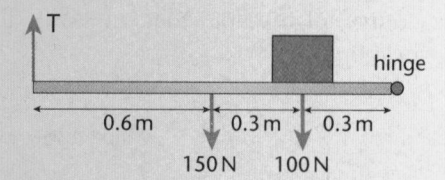

2. Explain how the hydraulic system shown in the diagram below multiplies the input force.

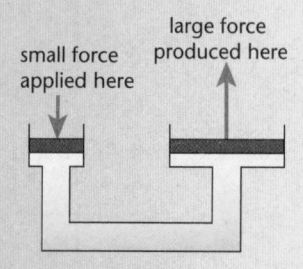

3. Describe how you would measure the density of air.

4. Use the principle of moments to calculate the value of force F_1 acting on the balanced beam in the diagram below.

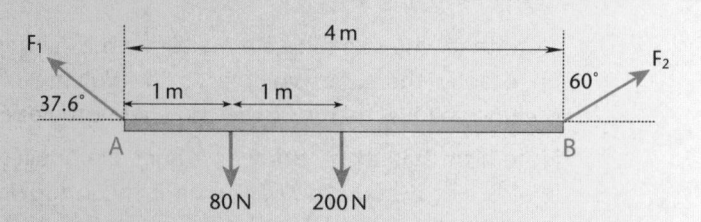

5. Why must hot air balloons be large?

1. 100 N
2. The input force acts on a small area, creating a large pressure. This pressure is transmitted through the liquid. The pressure then acts on a large area, creating a large force.
3. Weigh an air-filled container of known volume. Remove the air using a vacuum pump and weigh again. Subtract the mass of the empty container from that of the full container to find the mass of the air and divide this by the volume of the container.
4. 262 N
5. So that they displace a weight of air that is equal to their own average weight, including the load they carry. (The part of the balloon that is less dense than air must have a large volume so that the average density of the whole structure is the same as or less than the density of the air.)

1.7 Force and materials

After studying this section you should be able to:

- describe the difference between elastic and plastic behaviour
- state Hooke's law and appreciate its limitations
- contrast the behaviour of a ductile, a brittle and a polymeric solid when stretched
- explain how stress and strain are used in measurements of the Young modulus

LEARNING SUMMARY

Elastic or plastic

No material is absolutely rigid. Even a concrete floor changes shape as you walk across it. The behaviour of a material subjected to a tensile (pulling) or compressive (pushing) force can be described as either **elastic** or **plastic**.

> a material is **elastic** if it returns to its original shape and size when the force is removed
>
> a material is **plastic** if it does not return to its original shape and size when the force is removed

Most materials are elastic for a certain range of forces, up to the **elastic limit**, beyond which they are plastic. Plasticine and playdough are plastic for all forces.

Hooke attempted to write a simple rule that describes the behaviour of all materials subjected to a tensile force.

> If the extension is proportional to the stretching force, then doubling the force causes the extension to double.

KEY POINT

Hooke's law states that:

the extension, x, of a sample of material is proportional to the stretching force:
$$x \propto F$$
This can be reinterpreted as: $F = kx$
where k represents the stiffness of the sample and has units of $N\,m^{-1}$

> A polymeric solid is one made up of long chain molecules.

Metals and springs 'obey' Hooke's law up to a certain limit, called the **limit of proportionality**. For small extensions, the extension is proportional to the stretching force. Rubber and other **polymeric** solids do not show this pattern of behaviour.

The graphs below contrast the behaviour of different materials subjected to an increasing stretching force.

> Because the limits of proportionality and elasticity lie close together on the force–extension curve, the term 'elastic limit' is often used to refer to both points.

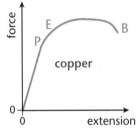
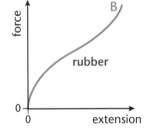
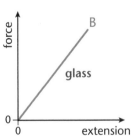

P = limit of proportionality   E = elastic limit   B = breaking point

- **Copper** is a **ductile** material, which means that it can be drawn into wires. It is also **malleable**, which means that it can be reshaped by hammering and bending without breaking. When stretched beyond the point E on the graph it retains its new shape.
- **Rubber** does not follow Hooke's law and it remains elastic until it breaks.
- **Glass** is **brittle**; it follows Hooke's law until it snaps.

Here are other properties of materials:

- **Kevlar** is **tough**; it can withstand shock and impact.
- **Mild steel** is **durable**; it can withstand repeated loading and unloading.
- **Diamond** is **hard**; it cannot be easily scratched.

Storing energy

The force–extension graph on the right shows that the more a material is stretched, the greater the force that is needed.

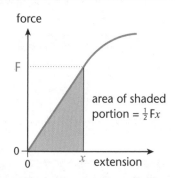

Section 1.4 (page 39) shows that work must be done when a force causes a displacement. So work must be done to stretch a material. Energy must be provided to the material to stretch it. You only have to pull on a stiff spring to experience this. That energy becomes available when the material returns to its original shape. You have to take care how you let go of a stretched stiff spring. So, a stretched material acts as an energy store. (Many non-electrical clocks, for example, use a spring as an energy store, without which they would stop quite quickly due to natural dissipation.)

The energy stored as a material is deformed and is represented by the area between the curve and the extension axis.

If a force F produces an extension x then the energy stored is equal to $\frac{1}{2}Fx$.

Since $F = kx$, energy stored $= \frac{1}{2}kx^2$.

Stress, strain and the Young Modulus

The extension that a force produces depends on the dimensions of the sample as well as the material that it is made from. To compare the elastic properties of materials in a way that does not depend on the sample tested, measurements of **stress** and **strain** are used instead of force and extension.

The stress on these wires is the same, even though the forces are different.

For some materials, e.g. copper and mild steel, the stress when the material breaks is less than the ultimate tensile stress.

The Young modulus for mild steel is 2.1×10^{11} Pa and that for copper is 0.7×10^{11} Pa. This means that for the same stretching force, a sample of copper stretches three times as much as one of mild steel with the same dimensions.

> **KEY POINT**
> stress = normal force per unit area
> $\sigma = F/A$
> Like pressure, stress is measured in Pa, where 1 Pa = 1 N m⁻²
> strain = extension per unit length
> $\varepsilon = x/l$
> Strain has no units.

The strength of a material is measured by its **breaking stress** or **ultimate tensile stress**. This is the maximum stress the material can withstand before it fractures or breaks and may not be the actual stress when breaking occurs. Its value is independent of the dimensions of the sample used for the test.

Using measurements of stress and strain, different materials can be compared by the **Young modulus**, E.

> **KEY POINT**
> Young modulus = stress ÷ strain
> $E = \sigma/\varepsilon = F/A \div x/l = Fl/xA$
> The Young modulus has the same units as stress, Pa.

The greater the value of the Young modulus the **stiffer** the material, i.e. the less it stretches for a given force.

Some classes of solid materials

OCR A   1
OCR B   1

Metals

Metals are defined by their chemical properties – they are elements whose atoms tend to lose electrons to become positive ions. Metals also have clear patterns of physical properties – they have dense crystalline structures, usually with high melting points, and can be thought of as having crystal lattices immersed in a 'sea' of mobile electrons. The presence of mobile electrons makes metals good conductors of electricity and of heat. Many metals are *ductile* and *malleable*, so they can be worked into useful shapes.

Ceramics

Ceramic materials are made by heating softer and often granular material, such as clay, to very high temperatures. The heating causes change of inter-atomic structure, producing material that is usually *brittle* but resistant to further high temperature and to chemical corrosion.

Polymers

Polymers consist of long chain molecules, which are ordered or disordered to different extents. Rubber has less ordered structure, polythene has more so. Cross-linking between the chains has a strong effect on polymer physical behaviour, making the materials much harder.

Composite materials

Composite materials can offer the benefits of the different materials that they are made up from. Carbon fibres embedded in polymer material, for example, combine the low density of the polymer with the strength of the carbon fibres. This produces materials that can replace metals. Sports such as tennis and pole-vaulting have been transformed by the uses of composite materials.

Progress check

1. Describe the difference between a *ductile* material and one that is *brittle*.

2. A force of 150 N applied to a spring causes an extension of 25 cm.
   a. Calculate the energy stored in the spring.
   b. What assumption is necessary to answer a?

3. The Young modulus for copper is 1.3×10^{11} Pa.

   A tensile force of 150 N is applied to a sample of length 1.50 m and cross-sectional area 0.40 mm^2. (1.0 mm^2 = 1.0×10^{-6} m^2).

   Assuming that the limit of proportionality is not exceeded, calculate the extension of the copper.

3 4.3 mm
   b the limit of proportionality has not been exceeded
2 a 18.75 J
1 A ductile material can be drawn into wires; a brittle material snaps easily.

Sample questions and model answers

1 In an emergency stop, a driver applies the brakes in a car travelling at $24\,\mathrm{m\,s^{-1}}$. The car stops after travelling $19.2\,\mathrm{m}$ while braking.

(a) Calculate the average acceleration of the car during braking. [3]

The final velocity, v, is implicit in the question, although not clearly stated. As the car brakes to a halt, the final velocity must be 0.

> The initial velocity, u, final velocity, v, and displacement, s, are known.
> The unknown quantity is the acceleration, a.
> The appropriate equation is $v^2 = u^2 + 2as$ 1 mark
> Rearrange to give $a = (v^2 - u^2)/2s$ 1 mark
> $= (0 - 24^2)/(2 \times 19.2)$
> $= -15\,\mathrm{m\,s^{-2}}$ 1 mark

Note that the acceleration is negative, as the speed in the direction of the velocity is decreasing.

(b) Sketch a graph that shows how the speed of the car changes during braking, assuming that the acceleration is uniform. Mark the value of the intercept on the time axis.

Show on your graph the feature that represents the distance travelled during braking. [3]

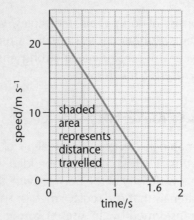

The graph shows a straight line from $24\,\mathrm{m\,s^{-1}}$ to $0\,\mathrm{m\,s^{-1}}$ 1 mark
The intercept on the time axis is $1.6\,\mathrm{s}$ 1 mark
The area between the graph line and the time axis represents the distance travelled.   1 mark

(c) Calculate the size of the force required to decelerate the driver, of mass $75\,\mathrm{kg}$, at the same rate as the car. [3]

Again, the negative sign shows that the direction of the force is opposite to the direction of motion.

> force $=$ mass $\times$ acceleration   1 mark
> $= 75\,\mathrm{kg} \times -15\,\mathrm{m\,s^{-2}}$ 1 mark
> $= -1125\,\mathrm{N}$ 1 mark

The cue word 'suggest' indicates that you are not meant to have studied this example as part of your course, but should be able to reason it from your understanding of forces and motion.

(d) Suggest what is likely to happen to a driver who is not wearing a seat belt. The car does not have air bags. [2]

> The driver will carry on moving (1 mark) as there is nothing to stop her/him until (s)he hits the steering wheel and windscreen (1 mark).

2 A wooden pole is held in an upright position by two wires in tension. These forces are shown in the diagram below.

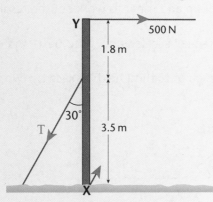

The pole is fixed at its base.

(a) Calculate the moment of the 500 N horizontal force about X.   [2]

moment = force x perpendicular distance to pivot
         = 500 N x 5.3 m   1 mark
         = 2650 N m   1 mark

(b) Calculate the value of the force T.   [3]

The horizontal component of the force T must also have a moment of
2650 N m: 1 mark

Tcos60° x 3.5 m = 2650 N m   1 mark

T = 2650 N m ÷ (cos60° x 3.5 m) = 1514 N   1 mark

Always write down the formula that you intend to use, and each step in the working. It would be very easy in this case to use a wrong distance. Simply writing down a wrong answer always gets no marks, but showing how you arrive at that wrong answer allows credit to be awarded for the correct process.

Only the horizontal component of T has a turning effect; its moment must counterbalance that of the 500 N horizontal force.

Having written down the physical principle, the next step is to write it in the form of an equation. Tcos60° or Tsin30° is the horizontal component of T.

The final step is to calculate the value of T. Although it is not essential to show the units of physical quantities at each step in the calculation, it is good practice and you MUST include the correct unit with your answer to each part.

This problem cannot be solved by resolving T and the 500 N force as there is a force at X (you can see that this is not just a vertical reaction force by taking moments about Y – there must be a horizontal component with a moment that balances the horizontal component of T).

Practice examination questions

1 A microlight aircraft heads due north at a speed through the air of $18\,\text{m s}^{-1}$.
A wind from the west causes the aircraft to move east at $6\,\text{m s}^{-1}$.
Draw a vector diagram and use it to work out the resultant velocity of
the aircraft. [3]
Take the value of free-fall acceleration, g, to be $10\,\text{m s}^{-2}$.

2 The diagram shows a light fitting held in place by two cables.

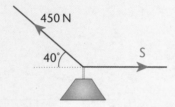

(a) Calculate the vertical component of the 450 N force. [2]

(b) Write down the weight of the light fitting. [1]

(c) Explain why a vector diagram that represents the forces on the light fitting
is a closed triangle. [2]

(d) Draw the vector diagram and use it to find the value of the force S. [2]

3 The diagram is a velocity–time graph for a train travelling between two stations.
A positive velocity represents motion in the forwards direction.

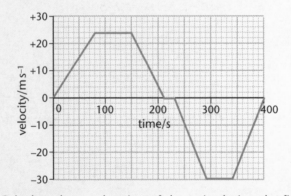

(a) Calculate the acceleration of the train during the first 80 s. [3]

(b) During which other times shown on the graph is the train increasing
speed? [1]

(c) For what length of the time was the train:
(i) not moving? [1]
(ii) travelling at a constant non-zero speed? [1]

(d) Calculate the total distance travelled by the train. [2]

(e) At the end of the time shown on the graph, how far was the train from
its starting point? [1]

4 In a game of cricket, a ball leaves a bat in a horizontal direction from a height
of 0.39 m above the ground.

The speed of the ball is 35 m s^{-1}.

(a) Calculate the time interval between the ball leaving the bat and reaching the ground. [3]

(b) What horizontal distance does the ball travel in this time? [2]

(c) Fielders are not allowed within 15 m of the bat.
What is the maximum speed that the ball should leave the bat, travelling in a horizontal direction, for the player not to be caught out? [2]

5 A motorist travelling at the legal speed limit of 28 m s^{-1} (60 mph) takes his foot off the accelerator as he passes a sign showing that the speed limit is reduced to 14 m s^{-1} (30 mph). The car decelerates at 2.0 m s^{-2}.

(a) For what time interval is the motorist exceeding the speed limit? [2]

(b) How far does the car travel in that time? [2]

6 A fountain is designed so that the water leaves the nozzle and rises vertically to a height of 3.5 m.

(a) Calculate the speed of the water as it leaves the nozzle. [3]

(b) For how long is each drop of water in the air? [2]

Take the value of free-fall acceleration, g to be 10 m s^{-2}.

7 An aircraft has a total mass, including fuel and passengers, of 70 000 kg. Its take-off speed is 60 m s^{-1} and it needs to reach that speed before the end of the runway, which is 1500 m long.

(a) Calculate the minimum acceleration of the aircraft. [3]

(b) Calculate the average force needed to achieve this acceleration. [3]

(c) Explain how the resultant force on the aircraft is likely to change during take-off. [2]

8 A car pulls a trailer with a force of 150 N. This is shown in the diagram.

(a) According to Newton's third law, forces exist in pairs.
  (i) What is the other force that makes up the pair? [1]
  (ii) Write down the size and direction of this force. [1]

(b) The mass of the trailer is 190 kg. As it sets off from rest, there is a resistive force of 20 N.
  (i) Calculate the size and direction of the resultant force on the trailer. [1]
  (ii) Calculate the initial acceleration of the trailer. [3]

(c) The force from the car on the trailer is maintained at 150 N.
When the car and trailer are travelling at constant speed:
  (i) what is the size of the resultant force on the trailer? [1]
  (ii) write down the size and direction of the resistive force on the trailer. [1]

Practice examination questions *(continued)*

9 The diagram shows the forces acting on a child on a playground slide. Air resistance is negligible.

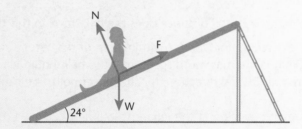

(a) The mass of the child is 40 kg.
  (i) Calculate the size of the force W. [1]
  (ii) Describe the force that, along with W, makes up the 'equal and opposite pair' of forces described by Newton's third law. [1]

(b) The size of the friction force is 90 N.
  (i) Calculate the component of W parallel to the slide. [2]
  (ii) Calculate the acceleration of the child. [3]

(c) The slide is 5.5 m long. If the child maintains this acceleration, calculate her speed at the bottom. [3]

10 An astronaut on a space walk pushes against the side of the spacecraft.

(a) Explain how this causes the velocity of both the astronaut and the spacecraft to change. [2]

(b) The astronaut can change his velocity by firing jets of nitrogen gas. Explain how this changes the velocity of the astronaut. [2]

11 A crane lifts a 3 tonne (3000 kg) load through a vertical height of 18 m in 24 s. Calculate:

(a) the work done on the load [3]

(b) the gain in gravitational potential energy of the load [1]

(c) the power output of the crane [2]

(d) the power input to the crane if the efficiency is 0.45. [2]

12 In a hydroelectric power station, water falls at the rate of 2.0×10^5 kg s^{-1} through a vertical height of 215 m before driving a turbine.

(a) Calculate the loss in gravitational potential energy of the water each second. [3]

(b) Assuming that all this energy is transferred to kinetic energy of the water, calculate the speed of the water as it enters the turbines. [3]

(c) The water leaves the turbines at a speed of 8 m s^{-1}. Calculate the maximum input power to the turbines. [2]

(d) The electrical output is 250 MW. Calculate the efficiency of this method of transferring gravitational potential energy to electricity. [2]

13 The diagram shows a child's toy. The spring is compressed a distance of 0.15 m using a force of 25 N. When the release mechanism is operated the ball rises into the air.

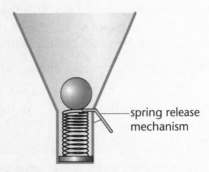

    spring release mechanism

(a) Calculate the energy stored in the spring. [2]

(b) The mass of the ball is 0.020 kg. Calculate:

    (i) the maximum speed of the ball [3]
    (ii) the maximum vertical distance that the ball travels. [3]

14 The diagram shows a garden tool being used to cut through a branch.

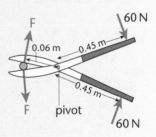

(a) Calculate the moment of the force on each handle. [2]

(b) Assuming equilibrium, calculate the value of the resistive force, F, that acts on each blade. [2]

(c) Suggest why a similar tool designed to cut through thicker branches has longer handles. [2]

15 The diagram shows the base of a steel girder that supports the roof of a building. The downward force in the girder is 4500 N and it is fixed to a square steel plate, each side of which is 0.30 m long.

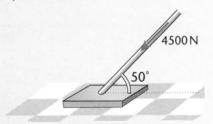

(a) Calculate the values of the horizontal and vertical components of the force in the girder. [2]

(b) Calculate the pressure that the steel plate exerts on the ground. [3]

Practice examination questions (continued)

(c) Explain why the girder is fixed to a steel plate rather than being fixed directly to the ground. [2]

16 A large aquarium has a horizontal cross-section of 10.5 m × 4.2 m. It is filled with water to a depth of 3.5 m. The density of water is 1000 kg m⁻³.

(a) Calculate the weight of the water in the aquarium. [3]

(b) Calculate the pressure on the base of the aquarium due to the weight of the water. [3]

(c) Explain why the glass used for the sides of the aquarium needs to be as strong as the base. [2]

(d) A similar aquarium has a horizontal cross-section of 10.5 m × 2.1 m and is filled with water to the same depth.
How does the pressure on the base compare to that calculated in (b)?
Give the reason for your answer. [2]

17 (a) How does a ductile material behave differently from a brittle one when subjected to a stretching force? [2]

(b) The graph shows the relationship between the stress on a sample of cast iron and the strain that it causes. The curve ends when the sample breaks.

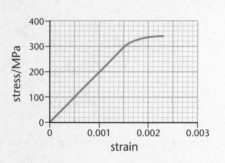

(i) Write down the value of the breaking stress of the cast iron. [1]
(ii) For what range of stresses does the cast iron follow Hooke's law? [1]
(iii) Calculate the Young modulus for the cast iron in the region where it follows Hooke's law. [2]
(iv) A second sample of the same material has twice the diameter of the original sample.
Explain whether the stress–strain graph for this sample would be the same as that shown in the diagram. [2]

Chapter 2

Electricity

The following topics are covered in this chapter:

- *Charge, current and energy transfer*
- *Resistance*
- *Circuits*

2.1 Charge, current and energy transfer

Charge

Charge is the name that we give to the ability of a body to exert and experience electrical force. The size of the force depends, among other things, on the amounts of charge involved.

It is useful to compare charge with mass. A body with mass can exert gravitational force on others that also have mass. The size of the force depends, among other things, on the amounts of mass involved.

There are differences as well as similarities. Gravitational forces act on huge scales and are most significant between large bodies. Electric forces act on smaller scales and they are most important when between small bodies. Also, gravitational forces are always attractive, while electric forces can be attractive or repulsive. So we have to suppose that while there is only one kind of mass, there are two kinds of charge. These need names – so we call them positive and negative. Charges that are the same always repel. Charges that are different always attract.

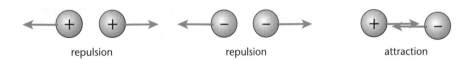

repulsion   repulsion   attraction

Electrons, protons and neutrons are particles which all have mass. Only two of them though, have electric charge. Protons have positive charge and electrons have negative charge. Neutrons get their name from their lack of electrical behaviour – they are electrically neutral.

Charge carriers

Within a metal, some of the electrons are free to move between individual atoms. The electrons act as **charge carriers** inside the metal. When they move together in large numbers there is a flow – there is an electric current.

An ordinary atom has a balance of electrons and protons, and is electrically neutral. However, one or more electrons can be removed from an atom, or extra electrons can be added. The atom then becomes an ion. Ions in solutions can be mobile. This happens in salt solutions, for example. The salt solution can conduct electricity thanks to the presence of these mobile charge carriers.

If a gas is ionised then it starts to conduct electricity. A flame, for example, contains ions, as does the gas inside a lighting tube.

Electric current

AQA A	1	Edexcel context	2
AQA B	2	OCR A	2
CCEA	1	OCR B	1
Edexcel concept	2	WJEC	1

The charge carried by one electron, $e = -1.60 \times 10^{-19}$ C. 1 C of charge is equivalent to that carried by 6.25×10^{18} electrons.

Electric current is 'rate of flow of charge'. Every word in the phrase counts – current is not just a flow of charge, but a 'rate of flow of charge'. The size of a current depends on how much charge is flowing and on how quickly. If ΔQ is amount of charge and Δt is the time it takes to flow past a point, then we can write an equation for current:

Current, $I = \Delta Q/\Delta t$

Current is measured in amperes or amps, or A for short.

We define the unit of charge in terms of the unit of current. The unit of charge is called the coulomb.

One coulomb is the amount of charge that flows past a point when a steady current of 1 A passes for 1 second.

Charging by friction

WJEC	1

Electrons cannot move around inside insulators. Every electron is attached to an atom. However, a tiny proportion of electrons can be transferred from one surface to another, simply by friction between them. The proportion of total electrons in a material that are moved from one to another in this way may be small, but the effects can be very noticeable.

A Van de Graaff generator system transfers electrons in large numbers, by a process that is a bit more than simple friction.

Current and drift velocity

AQA A	1	OCR A	2
AQA B	2	WJEC	1
Edexcel concept	2		
Edexcel context	2		

The free electrons in a metal are in constant random motion. A typical speed in a metal at room temperature is of the order of 1×10^6 m s^{-1}.

the random motion of a free electron, with collisions causing changes in direction

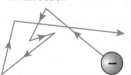

drift velocity is slow compared to the random motion of the electrons

Applying an electric force in the form of a voltage does not cause the free electrons to suddenly dash for the positive terminal. The random motion continues, but in addition the body of electrons moves at low speed in the direction negative to positive.

In a metal, a typical drift velocity is of the order of $1\,\text{mm}\,\text{s}^{-1}$, but in a semiconductor of similar dimensions carrying a similar current it is likely to be higher than this because the concentration of charge carriers is much less.

The overall speed of the body of electrons is called the drift velocity (V). There are three key variables that are related to the drift velocity:

- current (I)
- charge carrier concentration, n, defined as the number of charge carriers (free electrons in the case of a metal) per unit volume
- the cross-sectional area of the specimen, A.

These lead to the relationship for a metal $I = nAev$, where e is the electronic charge.

A similar expression applies to non-metallic conductors. As the charge on each ion can vary between non-metals, the expression becomes $I = nAqv$, where q is the ionic charge.

Charge and energy transfer

AQA A	1	Edexcel context	2
AQA B	2	OCR A	2
CCEA	1	OCR B	1
Edexcel concept	2	WJEC	1

Charge is not used up or changed into anything else as it moves round a circuit. It transfers energy. It gains energy from the source of electricity (mains, battery or power supply) and transfers this energy out to the surroundings by way of the circuit components as it flows through them.

In the case of a filament lamp, the energy transfer from the charge is due to collisions of the electrons moving through the filament. With a motor or other electromagnetic device the charge loses energy as it has to work against repulsive forces in moving through the armature.

There are two separate physical processes:

- energy transfer to the charge from the source
- energy transfer from the charge through components such as lamps and motors to the world outside the circuit.

In each case, rather than measure the energy transfer for a single charge carrier, the transfer to and from each coulomb of charge is measured.

Electromotive force and potential difference

Electromotive force, e.m.f., measures the energy transfer per unit charge from the source. It can be written as:

$$E = W_{source}/q$$

A 6V battery transfers 6J of energy to each coulomb of charge that it moves around a circuit.

Electromotive force is a mis-named quantity. It describes energy transfer rather than force.

> **KEY POINT**
>
> Electromotive force, symbol E, is defined as being the energy transfer from the source in driving unit charge round a complete circuit including through the source itself. It is measured in volts (V), where 1 volt = 1 joule/coulomb (1 J C^{-1}).

Potential difference, p.d., between the terminals of a component measures the work done per unit charge as it flows through the component. It can be written as:

$$V = W_{component}/q$$

> **KEY POINT**
>
> The potential difference between two points, symbol V, is defined as being the energy transfer per unit charge from the charge to the circuit between the two points. Like e.m.f., it is measured in V.

- Electromotive force and potential difference are both defined in terms of the work done per unit charge, a key difference being whether the work is done on the charge or by the charge.
- The unit of e.m.f. and p.d. is the volt. *One volt is the potential difference between two points if 1 J of energy is transferred when 1 C of charge passes between the points.*

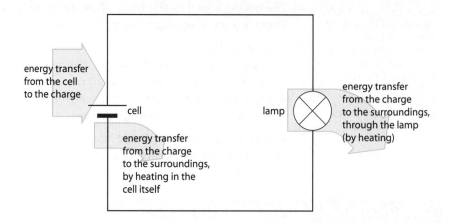

energy transfer from the cell to the charge

cell

energy transfer from the charge to the surroundings, by heating in the cell itself

lamp

energy transfer from the charge to the surroundings, through the lamp (by heating)

For this circuit:

$$\text{energy in} = \text{energy out}$$

$$\begin{array}{c}\text{energy transfer from}\\\text{the cell to the charge}\end{array} = \begin{array}{c}\text{energy transfer from}\\\text{the charge to the}\\\text{surroundings, through}\\\text{the lamp}\end{array} + \begin{array}{c}\text{energy transfer from}\\\text{the charge to the}\\\text{surroundings, by}\\\text{heating in the cell itself}\end{array}$$

The same charge flows through all parts of the circuit, so the whole equation can be divided by charge. That gives:

$$\text{e.m.f. of cell, } E = \text{p.d. across the lamp, } V + \begin{array}{c}\text{energy lost by cell per}\\\text{unit charge, also known}\\\text{as 'lost volts'}\end{array}$$

$$E = V + \text{lost volts}$$

Progress check

Use the definitions of current and potential difference.

1 Charge flows through a lamp filament at the rate of 15 C in 60 s. In the same time, 1500 J of energy is transferred to the filament.
Calculate:
a the current in the filament
b the potential difference across the filament.

2 A silicon diode has a cross-sectional area of 4.0 mm² (4.0 × 10⁻⁶ m²).
The concentration of charge carriers is 2.5×10^{27} m⁻³ and each charge has a charge of 1.60×10^{-19} C.
Calculate the drift velocity when the current in the diode is 1.5 A.

2 9.4 × 10⁻⁴ m s⁻¹
1 a 0.25 A   b 100 V

2.2 Resistance

After studying this section you should be able to:

- recall and use the formula for resistance
- recall and use the formula for resistivity
- state Ohm's law and describe its limitations
- sketch the current–voltage characteristics of a filament lamp, a wire at constant temperature and a diode
- explain how resistance varies with temperature for metals and semiconductors, and how this can be used for environmental sensing
- distinguish between resistance and conductance, and between resistivity and conductivity
- describe superconductivity
- predict energy transfer from values of quantities selected from current, voltage resistance and time
- predict the heating power of a resistor from values of quantities selected from current, voltage and resistance

LEARNING SUMMARY

Resistance

AQA A	1	Edexcel context	2
AQA B	2	OCR A	2
CCEA	1	OCR B	1
Edexcel concept	2	WJEC	1

Resistance is a measure of the opposition to current. The greater the resistance of a component, the smaller the current that passes for a given voltage.

> Resistance is defined and calculated using the formula:
> $$Resistance = voltage \div current \quad or \quad R = V \div I$$
> The unit of resistance is the ohm (Ω).
>
> **KEY POINT**

Different components

AQA A	1	Edexcel context	2
AQA B	2	OCR A	2
CCEA	1	OCR B	1
Edexcel concept	2	WJEC	1

It is possible to vary the potential difference between the terminals of a circuit component. To put it another way, it is possible to vary the voltage applied to the component. Then the current that results from different applied voltages can be measured. Different components behave differently, and the patterns of behaviour become easier to see, and then to explain, if the data is represented on graphs.

The graphs represent typical results obtained for a metal wire kept at a constant temperature, a filament lamp and a diode.

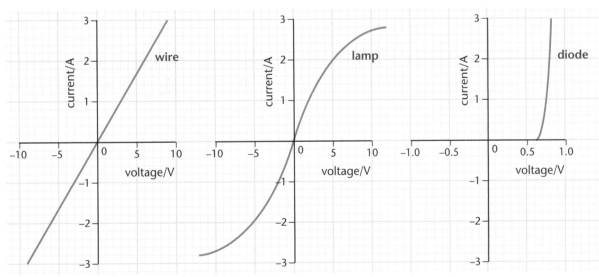

The text in *italics* is **Ohm's law**, one that he formulated based on his own experiments. Because it only applies to metallic conductors kept at a constant temperature, it is of only limited use in predicting the values of electric currents.

> **KEY POINT**
> The current–voltage graph for the **wire** is a straight line through the origin, showing that the *current is directly proportional to the voltage.*

A **filament lamp** is an everyday example of a metallic conductor that changes in temperature during use. In a short time after it is switched on, the temperature of the filament increases from room temperature to about 2000°C. The current–voltage graph shows how this change in temperature affects the relationship between these quantities. Like the one for the wire at constant temperature, this graph is symmetrical. Both the wire and the lamp filament show the same pattern of behaviour no matter which way round the voltage is applied.

Calculations based on this graph show that **as the current in the filament increases, so does its resistance**. This is due to the higher temperature causing an increase in the amplitude of the vibrations, which in turn increases the frequency of the collisions that impede the electron flow.

The arrow shows the direction in which the diode allows conventional current to pass.

The graph showing the characteristics of the **diode** also shows the property that makes diodes different from other components; they only allow current to pass in one direction.

Resistivity

AQA A	1	Edexcel context	2
AQA B	2	OCR A	2
CCEA	1	OCR B	1
Edexcel concept	2	WJEC	1

What factors determine the resistance of a resistor?
- Length
  Long wires have more resistance than short wires, so the longer the sample of material, the greater the resistance.

- Cross-sectional area
  Doubling the cross-sectional area of a sample also doubles the number of charge-carriers available to carry the current, halving the resistance.

> **KEY POINT**
> The resistance of a sample of material is directly proportional to its length and inversely proportional to its cross-sectional area.
> This statement can be written as R α l/A.

Good conductors have a low resistivity; that of copper being $1.7 \times 10^{-8} \,\Omega$ m. The resistivity of carbon is about one thousand times as great, $3.0 \times 10^{-5} \,\Omega$ m.

The other factor that determines the resistance is the material that the resistor is made of. The resistive properties of a material are measured by its **resistivity**, symbol ρ. When this is taken into account the formula becomes

$$R = \frac{\rho l}{A}.$$

This formula can be used to calculate the resistance when the dimensions and material of a resistor are known or to calculate the dimensions required to give a particular value of resistance.

> **KEY POINT**
> Resistance is a property of an individual component in a circuit, but resistivity is a property of a material. It is measured in units of Ω m.

Resistivity and temperature

NTC stands for negative temperature coefficient, indicating that its resistance decreases as temperature increases, and that the gradient of the resistance–temperature graph is negative.

Because of their large change in resistance with temperature, thermistors are useful in applications such as heat-sensitive cameras, which need to be able to detect small changes in temperature.

These graphs show the variation in resistance with environmental conditions for a metallic conductor, a thermistor and a light-dependent resistor.

The resistivity of a material is temperature dependent. Metals and semiconductors such as graphite (carbon) and silicon behave differently.

For metals, resistivity increases with temperature. At higher temperatures, a flow of electrons through a metal experiences more collisions due to atomic vibration, so that electrons tend to lose energy more quickly.

For semiconductors, the higher temperature sets more electrons (charge carriers) free within the material. So the resistivity of a semiconductor can decrease rapidly when its temperature increases.

An NTC thermistor is a sample of semiconductor, with suitable connections, that can be used in a circuit. The resistance of such a thermistor is very temperature-dependent, so it can allow an electric circuit to 'respond' to temperature changes.

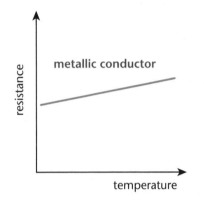

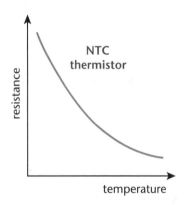

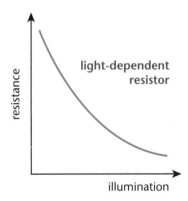

A light-dependent resistor, or LDR behaves in a similar way to a NTC thermistor, though it is light falling on a wafer of semiconductor that provides energy that sets electrons free. Again, the more free charge carriers there are per unit volume of a sample of material, the lower its resistance.

Conductance and conductivity

It is possible to consider the conductance of a component rather than its resistance. Conductance is the 'inverse' of resistance.

conductance = 1 / resistance

Conductance can be defined by the equation:

conductance = I/V and is measured in A V^{-1}.

In a similar way, conductivity is the inverse of resistivity. Conductivity is calculated from:

conductivity, $\sigma = l$ / R A

Superconductivity

The resistivity of some (but not all) metals falls to zero (which means that conductivity becomes infinite) at extremely low temperatures. Superconducting cables can carry very large currents without any energy loss to the surroundings. There is no dissipation, and no potential difference is needed to keep the current going.

Unfortunately, keeping the temperature of a cable very low is difficult and expensive. Superconducting cables are only used in applications where a strong magnetic field is needed, and where people are prepared to pay. These include medical imaging equipment (MRI scanners) and scientific research (such as in particle accelerators).

Some more complex combinations of materials become superconducting at higher, but still very cold, temperatures. These are called high temperature superconductors, but the 'high' is relative. Existing materials still need to be well over 100 °C below the freezing point of water before they become superconducting.

Energy transfer by resistors

Resistors are heated by the current within them. Electrons must 'do work' to overcome resistance, just as a car must do work to overcome resistance of different kinds (friction and air resistance). The amount of work done is exactly equal to the amount of energy that is transferred by heating.

energy transferred by heating, ΔE = work done by electrons, $\Delta W = Vq$

where V is the potential difference between the ends of the resistor and q is the charge carried by the electrons. Note that this equation comes from the definition of potential difference, found on page 69.

Charge q is related to the current:

$$I = q/\Delta t$$

Which is the same as saying that:

$$q = I\,\Delta t$$

So, during a length of time Δt, energy transferred by heating:

$$\Delta E = V\,I\,\Delta t$$

In a resistor, V = IR, so we could write,

$$\Delta E = I^2\,R\,\Delta t$$

And also, I = V/R, so: $\qquad\qquad \Delta E = V^2\,\Delta t\ /\ R$

Power

Unless stated otherwise, the term 'rate of' refers to 'rate of change with time', so power is the work done or energy transfer per second.

Power is the *rate of working or energy transfer*.

> **KEY POINT**
> The relationship between electrical power, current and potential difference is:
> *power = current × potential difference* or P = IV.
> Power is measured in watts (W), where 1 W = 1 J s⁻¹.

This relationship can be combined with the resistance equation to give:

- $P = I^2R$
- $P = V^2/R$

These three equivalent relationships can all be used to calculate the value of either a constant or a varying power.

A battery-operated lamp transfers energy at a constant rate. The power of a mains-operated lamp is constantly changing. The graphs compare the power of a battery-operated lamp with that of a mains lamp.

The smaller the value of Δt, the better the approximation that E = IVΔt is to the area between the graph line and the time axis.

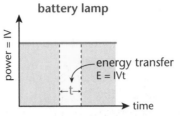

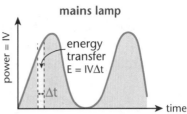

The total energy transfer in time Δt where the power is constant is equal to:

The SI unit of energy is the joule. For domestic purposes, energy is measured in kilowatt-hours (kW h), where 1 kW h is the energy transfer by a 1 kW appliance in 1 hour.

$$\Delta E = \text{power} \times \text{time} = IV\Delta t.$$

KEY POINT

For a varying power the energy transfer is represented by the area between the curve and the time axis of a graph of power against time. The shorter the value of Δt, the better the approximation of the above expression to this area.

Progress check

1 Use the resistance formula to work out the following:
   a The voltage across a lamp filament if the current passing is 2.4 A and the filament resistance is 7.3 Ω.
   b The resistance of an infra-red heater if the current is 9.5 A when it operates from the mains 240 V supply.
   c The current passing in a 6.8 Ω resistor when it is connected to a 9 V battery.

Note that 1 mm = 1 × 10⁻³ m.

2 Use the resistivity formula to work out the following:
   a The resistivity of silicon. A 1 mm cube of silicon has a resistance of 2.20 × 10⁶ Ω when measured across opposite faces.
   b The length of carbon rod required to make a 3.9 Ω resistor.
      The resistivity of carbon is 3.00 × 10⁻⁵ Ωm and the rod has a cross-sectional area of 1.26 × 10⁻⁶ m².

First re-arrange the resistivity equation to make *l* the subject.

3 An electrical heating element is to be made from nichrome wire. The current in the heating element is to be 5.00 A when the voltage across it is 12.0 V.
   a Calculate the resistance of the heating element.
      The resistivity of nichrome is 1.10 × 10⁻⁶ Ωm.
      The cross-sectional area of the wire used is 7.85 × 10⁻⁷ m².
   b Calculate the length of wire required.

4 What happens to the power of a filament lamp when the current is doubled?

5 When the current in a cell is 1.5 A it supplies energy at the rate of 18 W. Calculate the e.m.f. of the cell and the total resistance of the circuit.

6 a Show, using appropriate equations, what is meant by the statement 'conductivity is the inverse of resistivity'.
   b What are the units of conductivity?

1 a 17.5 V   b 25.3 Ω   c 1.32 A
2 a 2200 Ω m   b 0.164 m
3 a 2.4 Ω   b 1.71 m
4 It is quadrupled.
5 12 V and 8 Ω.
6 a σ = 1/ρ = l/RA,
   b Ω⁻¹ m⁻¹

2.3 Circuits

After studying this section you should be able to:

- calculate the effective resistance of a number of resistors connected in series or parallel
- understand that a source of electricity may contribute to the resistance of a circuit and the effects of this
- apply the principles of conservation of charge and conservation of energy to circuit calculations
- describe how the resistance of a metallic conductor, a thermistor and a light-dependent resistor change with environmental conditions

LEARNING SUMMARY

Series and parallel

AQA A		Edexcel context	2
AQA B	1	OCR A	2
CCEA	2	OCR B	1
Edexcel concept	2	WJEC	2

Resistors can be combined in series, parallel or a combination of the two. There are simple formulae for calculating the effective resistance of two or more resistors connected together.

> The effective resistance of a number of resistors in series is always greater than the largest value resistor in the combination.

Series

Adding another resistor in series in a circuit always decreases the current. This means that the effective resistance in the circuit has increased.

> **KEY POINT**
> The formula for calculating the effective resistance of a number of resistors in series is:
> $$R = R_1 + R_2 + R_3.$$

This formula can be used for any number of resistors connected in series by adding on extra terms.

> To find the effective resistance of a 4.7 Ω, a 6.8 Ω and a 10 Ω resistor connected in parallel
> $1/R = 1/4.7 + 1/6.8 + 1/10$
> $= 0.460$
> $R = 1/0.460 = 2.2\ \Omega$
> *A common error is to forget to carry out the final stage of the calculation.*

Parallel

When an additional resistor is added in parallel to a circuit, the current always increases. The additional resistor opens up another current path, without affecting the current in any of the existing paths. This results in less resistance in the circuit.

> The effective resistance of a number of resistors in parallel is always smaller than the smallest value resistor in the combination.

> **KEY POINT**
> The formula for calculating the effective resistance of a number of resistors in parallel is:
> $$1/R = 1/R_1 + 1/R_2 + 1/R_3.$$

As with the formula for resistors in series, any number of terms can be added to this expression.

Internal resistance of a cell

> The internal resistance of a cell or power supply acts as an extra resistor in series with the rest of the circuit.

All parts of a circuit have resistance and energy is needed to move the charge carriers through them. A cell of e.m.f. 1.5 V, for example, does 1.5 J of work on each coulomb of charge that completes a circuit.

This work is done on moving the charge through:

- the connecting wires
- the circuit components
- the cell itself, due to the cell's own **internal resistance**, symbol r.

The greater the current in the cell, the more work is done against the internal resistance, so the less can be done in the external circuit.

How does internal resistance affect the circuit?

For example, a cell of e.m.f (E) 1.5V and internal resistance (r) 0.2Ω causes a current (I) of 0.5A in an external resistor (R). The work done per coulomb of charge against the internal resistance = I × r = 0.5A × 0.2Ω = 0.1V. This leaves 1.4V for the external circuit.

> The greater the current in the cell, the greater the reduction in potential difference.

The effect of internal resistance is to reduce the potential difference across the external circuit.

When the only component connected to a cell is a high-resistance voltmeter such as a digital voltmeter, the reading is the cell's e.m.f. This is because when it is connected to the voltmeter alone, virtually no current passes in the cell so no work is being done against internal resistance. However, when the cell is connected to a circuit, a voltmeter placed across its terminals reads less than the e.m.f.

> A typical digital voltmeter has a resistance of 10 MΩ (1.0 x 10⁷ Ω), so the current passing in it is very small, causing negligible reduction in the p.d. across the terminals.

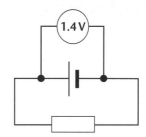

The high resistance voltmeter shows that the e.m.f. of the cell is 1.5 V. The cell supplies 1.5 J of energy to move each coulomb of charge around a complete circuit.

0.1 J of energy is needed to move each coulomb of charge through the cell, leaving 1.4 J to move the charge round the rest of the circuit.

> **KEY POINT**
>
> The relationship between e.m.f. and terminal p.d. is:
> e.m.f. = terminal p.d. + 'lost volts'
> E = V + Ir

Compare this equation with that on pages 69–70.

Kirchhoff's laws

In a circuit, the flow of charge transfers energy from the source to the components. Neither of these physical quantities becomes used up in the process. Both are conserved. Kirchhoff's laws are re-statements of these fundamental laws in terms that apply to electric circuits.

> Mass, charge and energy are fundamental conserved quantities. They cannot be created or destroyed.

> **KEY POINT**
>
> Kirchhoff's first law states that the total current that enters a junction is equal to the total current that leaves the junction.

This is illustrated in the diagram below and is simply a statement that charge is conserved at the junction.

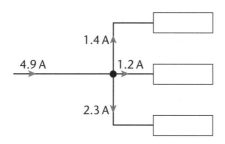

KEY POINT

Kirchhoff's second law is about conservation of energy. It states that around any closed loop (i.e. complete series path), the total e.m.f. is equal to the sum of the potential differences, $E = \Sigma(IR)$.

Note that R here is the total circuit resistance including the internal resistance of the source.

Remember that e.m.f. relates to energy transfer **to** the charge and p.d. relates to energy transfer **from** the charge, so this law is stating that when a quantity of charge makes a complete circuit, the energy transfers to the charge and from the charge are the same. The second law can be applied to a series circuit or any series path within a parallel circuit.

The variable resistor

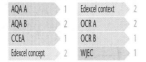

AQA A	1	Edexcel context	2
AQA B	2	OCR A	2
CCEA	1	OCR B	1
Edexcel concept	2	WJEC	1

A variable resistor allows the length of wire through which current must flow to be varied.

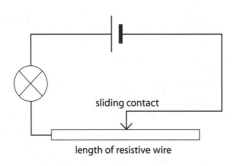

sliding contact

length of resistive wire

The potential divider

AQA A	1	Edexcel context	2
AQA B	2	OCR A	2
CCEA	1	OCR B	1
Edexcel concept	2	WJEC	1

A potential difference can be applied to a pair of resistors. When a current flows, provided that there are no junctions between the resistors, the current is the same in them both.

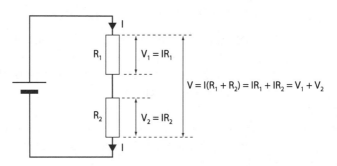

Remember: Potential difference is a DIFFERENCE between two points in a circuit. Here, the two differences, V_1 and V_2, one after the other, add up to the total difference, V.

V is the total potential difference across both resistors. V_1 and V_2 are the p.d.s between the ends of resistor R_1 and resistor R_2.

$$V = V_1 + V_2$$

V_1 and V_2 will only be the same if the two resistances are equal. If one resistance is bigger than the other then it will have the bigger potential difference, as is suggested by these two equations:

$$V_1 = IR_1 \qquad\qquad V_2 = IR_2$$

The system is called a **potential divider**. Connections to the two ends of one of the resistors can be used to apply a potential difference to further components. These can be called output devices or components.

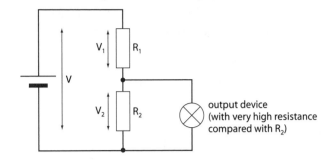

The potential difference between the ends of the lamp is the same as that between the ends of resistor R_2.

Input and output voltage

The total potential difference applied to the two resistors of a potential divider system can be called the input voltage. The potential difference that is then applied to an output device is called the output voltage.

If one of the resistors in a potential divider is an LDR, for example, then changes in light intensity cause its resistance to change. This causes the output voltage to change. This means that the behaviour of the output components changes when light intensity changes.

Thermistors and other 'environmental sensing' devices can be used in place of, or even as well as, LDRs. The behaviour of the output component then depends on temperature. So potential dividers can form the basis of 'sensing circuits'.

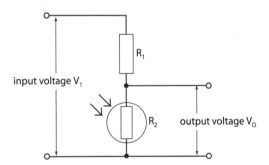

In an investigation in which potential difference is controlled (as the independent variable), a continuous wire with a sliding connection can act as a potential divider. When the slider is at one end, the 'output' voltage is equal to the total potential difference applied to the wire. When the slider is at the other end then the output voltage is zero.

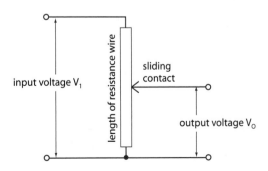

There is an equation that can be used as a predictive tool, so that it is possible to know what values of resistor to use with a particular input voltage, in order to achieve a required output voltage.

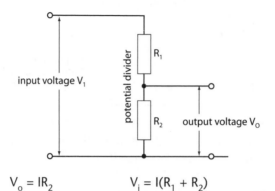

$$V_o = IR_2 \qquad\qquad V_i = I(R_1 + R_2)$$

These equations together tell us that $\dfrac{V_o}{V_i}$ and $\dfrac{IR_2}{I(R_1 + R_2)}$ are identical.

$$\frac{V_o}{V_i} = \frac{IR_2}{I(R_1 + R_2)} = \frac{R_2}{(R_1 + R_2)}$$

So, $V_o = \dfrac{R_2}{R_1 + R_2} V_i$

Using an oscilloscope to measure voltage

AQA A   1

Potential difference results in a force acting on charged bodies. If a potential difference is applied to two metal plates then electrons between the plates experience force. If they are moving in a beam then the beam is deflected. The amount of deflection depends on the size of the potential difference. This is what happens in an oscilloscope, which has two parallel horizontal metal plates connected to input terminals. A beam of electrons hits the screen to create a glowing dot.

The oscilloscope can be set to produce a single dot at the centre of the screen. When a voltage is then applied to the input terminals and so to the plates, the beam is deflected vertically and the dot moves.

The amount of deflection provides a measure of voltage, and the screen has a grid to make this easier. The sensitivity can be varied.

The oscilloscope also has a time base control. Switching the time base on applies a varying potential difference to two vertical plates and this makes the dot scan horizontally across the screen. Most scanning speeds are fast, so that the dot appears as a single horizontal straight line.

These diagrams compare the traces obtained for a 3.0 V direct voltage input with different sensitivity and time base settings.

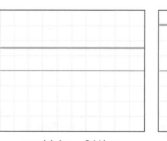

sensitivity = 2 V/cm
time base on

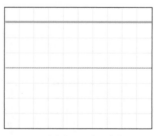

sensitivity = 1 V/cm
time base on

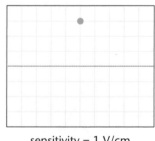

sensitivity = 1 V/cm
time base off

Using an oscilloscope to measure time and frequency

AQA A   1

If the voltage varies, the dot moves up and down. If the time base is switched on and set to a particular sensitivity the horizontal distance across the screen acts rather like a time axis.

If the voltage that is applied to the oscilloscope input terminals varies periodically, in a regular pattern, then the time for one cycle can be read from the screen, provided that the setting of the time base control is known. From the time for one cycle, or period, the frequency can be calculated.

- A time interval is measured by multiplying the appropriate horizontal distance on the screen by the time base setting.
- A frequency is measured by calculating (1/time) for one complete cycle.

In the example shown in the diagram, the time base is set to 1 ms cm⁻¹, or 1.0×10^{-3} s cm⁻¹.

Remember to convert times in ms or μs into s before calculating a frequency.

One complete cycle of the wave occupies a distance of 4.0 cm on the screen, so the time taken for one cycle is 4.0×10^{-3} s.

Since *frequency* = 1/*time period* or f = 1/t, the frequency of the wave is $1 \div (4.0 \times 10^{-3})$ s = 250 Hz.

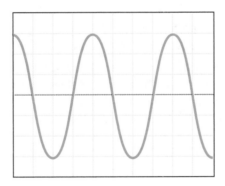

Progress check

1 Calculate the effective resistance of each of the following combinations of resistors.

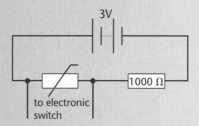

In **b**, work out the resistance of the parallel combination first.

2 A cell has an e.m.f. of 1.6 V and internal resistance 0.5 Ω.
Calculate the current in the cell and the p.d. across the terminals when it is connected to:
  a a 0.5 Ω resistor
  b a 3.5 Ω resistor.

Remember that the p.d. across the terminals is equal to that across the external resistor.

3 A battery of e.m.f 4.5 V and internal resistance 0.6 Ω is connected to a 3.3 Ω resistor in series with a 5.1 Ω resistor.
Calculate:
  a the current in the circuit
  b the p.d. across each resistor
  c the p.d. across the terminals of the cell.

4 The diagram shows a thermistor used in a potential divider as part of a temperature-controlled switch. The internal resistance of the power supply is small and can be neglected.

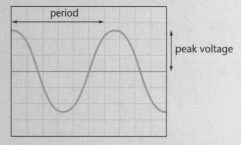

The electronic switch turns on a heater when the p.d. across the thermistor rises to 0.6 V.

The quickest way to answer **a** is to use ratios.

  a Calculate the p.d. across the thermistor when its resistance is 200 Ω.
  b Explain how the heater becomes switched off.
  c Calculate the resistance of the thermistor when the heater switches off.

5 The diagram shows an oscilloscope display of an alternating voltage.

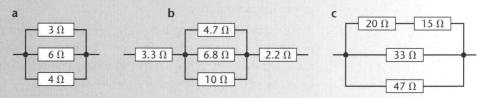

The sensitivity is set to 10 V/cm and the time base to 5 ms/cm. Calculate:
  a the peak value of the voltage
  b the frequency of the voltage.

Sample question and model answer

(a) What is the unit of resistivity? [1]

Since $\rho = AR/l$, the units of ρ must equal those of AR/l.
Units of $AR/l = m^2 \times \Omega \div m = \Omega m$.   1 mark

This technique relies on the fact that a relationship between physical quantities must be homogeneous, i.e. the units on each side are the same.

(b) A cable consists of seven straight strands of copper wire each of diameter 1.35 mm as shown in the diagram. [6]

strand of copper wire

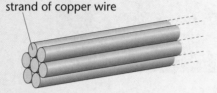

Calculate:
(i) the cross-sectional area of **one strand** of copper wire.

Cross-sectional area $= \frac{1}{4}\pi d^2$
$= \frac{1}{4} \times \pi \times (1.35 \times 10^{-3})^2 = 1.43 \times 10^{-6}\ m^2$ 1 mark

It is important here to convert the diameter from mm to m in order to maintain a consistent set of units. Many AS candidates, faced with this type of calculation, work out the answer in mm² and then apply a wrong conversion factor to change it to m².

(ii) the resistance of a 100 m length of the cable, given that the resistivity of copper is $1.6 \times 10^{-8}\ \Omega$ m.

The resistance of one strand:
$R = \rho l / A$ 1 mark
$= 1.6 \times 10^{-8}\ \Omega m \times 100 m \div (1.43 \times 10^{-6})m^2 = 1.12\ \Omega$ 1 mark
The resistance of 100 m of the cable is $\frac{1}{7}$ of this, i.e. $0.16\ \Omega$.   1 mark

The emphasis here is on reading the question thoroughly. A common error is to calculate the resistance of one strand only, ignoring the fact that the resistance of the cable is emphasised in the question.

(iii) The cable carries a current of 20 A. What is the potential difference between the ends of the cable? [2]

$V = IR = 20 A \times 0.16\ \Omega = 3.2 V$.   1 mark

The calculations in (c) are straightforward applications of the resistance equation. Do not worry if your answers to (b) were wrong. You still gain full marks in (c) for correct working with the wrong resistance values.

(iv) If a single strand of the copper wire in part (b) carried a current of 20 A, what would be the potential difference between its ends?

$20 A \times 1.12\ \Omega = 22.4 V$.   1 mark

Practice examination questions

1 (a) Write down the formula that relates the resistivity of a material to the resistance of a particular sample. [1]

(b) Explain the difference between *resistance* and *resistivity*. [2]

(c) The diagram shows a cuboid made from carbon.

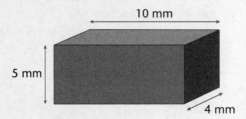

10 mm

5 mm

4 mm

The resistivity of carbon is 3.00×10^{-5} Ωm.
Calculate the resistance of the cuboid, measured between the faces that are 10 mm apart. [3]

2 In the circuit shown in the diagram, the internal resistance of the cell can be neglected.

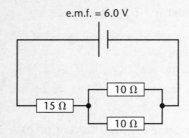

e.m.f. = 6.0 V

15 Ω

10 Ω

10 Ω

Calculate:
(a) the effective resistance of the two resistors in parallel [2]

(b) the current in the 15 Ω resistor [3]

(c) the current in each 10 Ω resistor [2]

(d) the potential difference across each 10 Ω resistor. [2]

3 (a) (i) Draw a sketch graph to show how the resistance of a thermistor varies with increasing temperature. [2]

(ii) State TWO ways in which this graph differs from a graph that shows how the resistance of a metallic conductor varies with increasing temperature. [2]

(iii) Explain why Ohm's law does not apply to either graph. [2]

(b) The diagram shows a thermistor in a series circuit with a 100 Ω resistor.

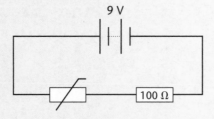

9 V

100 Ω

Practice examination questions (continued)

The table gives some data about the resistance of the thermistor.

Temperature/°C	Thermistor resistance/Ω
5	500
20	300
40	75

(i)   Calculate the potential difference across the thermistor at a temperature of 5°C.   [3]

(ii) Explain how the potential difference across the thermistor changes as the temperature rises.   [2]

(iii) The potential difference across the resistor is used to switch off a heater when the temperature reaches 40°C. Calculate the potential difference at which the switch operates.   [3]

(iv) The circuit is adapted to switch off the heater when the temperature reaches 20°C. Calculate the value of the fixed resistor required.   [3]

4   A cell has an e.m.f. of 1.60 V. When connected to a digital voltmeter of resistance 10 MΩ ($1.0 \times 10^7 \Omega$), the reading on the voltmeter is 1.60 V. When the cell is connected to a moving coil voltmeter of resistance 1 kΩ ($1.0 \times 10^3 \Omega$) the reading on the voltmeter is 1.55 V.

(a) Explain why there is a difference in the voltmeter readings.   [3]

(b) Calculate the internal resistance of the cell.   [3]

5   A 12 V car battery is used to light four 6 W parking lamps connected in parallel.

(a) Calculate the current in the battery.   [1]

(b) How much charge flows through the battery in one minute?   [3]

(c) Calculate the effective resistance of the four lamps.   [3]

6   (a) A 4.7 Ω resistor has a power rating of 10 W.

   (i)   Calculate the maximum current that should pass in the resistor.   [3]

   (ii) Calculate the maximum potential difference across the resistor.   [2]

   (iii) The resistor is manufactured from constantan wire of cross-section 2.0×10^{-7} m^2 and resistivity $5.0 \times 10^{-7} \Omega$ m.
         Calculate the length of wire used to make the resistor.   [3]

   (b) Low-power resistors are often made from carbon. The resistivity of carbon is $3.0 \times 10^{-5} \Omega$ m. What is the advantage of making resistors from carbon?   [2]

Practice examination questions (continued)

7 The diagrams show two circuits that can be used to investigate the current–voltage characteristics of component X.

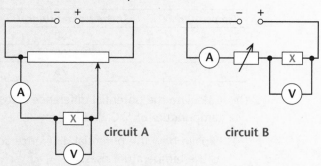

circuit A   circuit B

(a) What name is given to circuit A?   [1]

(b) Describe what circuit A is able to do.   [1]

(c) What is the advantage of using circuit A rather than circuit B to investigate the characteristics of a diode?   [2]

(d) The table shows data obtained by varying the voltage across a diode.

voltage/V	0.65	0.71	0.76	0.78	0.80
current/A	0.06	0.20	0.44	0.68	1.00
resistance/Ω					

(i) Complete the table.   [2]

(ii) Draw a graph of resistance against voltage.   [4]

(iii) Use the graph to estimate the voltage at which the resistance of the diode is $1\,\Omega$.   [1]

(iv) Suggest why the data only span a narrow voltage range.   [1]

8 In a simple lighting circuit, a 60 W lamp is connected to a 12 V battery using copper cable of cross-section $1.25\,mm^2$ ($1.25 \times 10^{-6}\,m^2$). The lamp filament is made of tungsten of cross-section $3.0 \times 10^{-4}\,mm^2$ ($3.0 \times 10^{-10}\,m^2$). The number of free electrons per unit volume for copper is $8.0 \times 10^{28}\,m^{-3}$ and that for tungsten is $3.4 \times 10^{28}\,m^{-3}$. The electronic charge $e = -1.60 \times 10^{-19}\,C$.

(a) Calculate the drift velocity of the electrons in the copper cable.   [3]

(b) Give TWO reasons why the drift velocity of the electrons in the tungsten is much greater than that of electrons in the copper.   [2]

(c) Suggest why the tungsten becomes heated while the copper remains cool.   [2]

9 A cell of e.m.f. 3.0 V and internal resistance $0.5\,\Omega$ is connected to a $3.0\,\Omega$ resistor placed in series with a parallel combination of a $6.0\,\Omega$ and a $12.0\,\Omega$ resistor.

Calculate:

(a) the current in the circuit   [3]

(b) the p.d. across the terminals of the cell.   [2]

Chapter 3

Waves, imaging and information

The following topics are covered in this chapter:

- *Wave behaviours*
- *Sound and superposition*
- *Light and imaging*
- *Information transfer*

3.1 Wave behaviours

After studying this section you should be able to:

- *explain that waves transmit energy and information from point to point without transmitting energy between the two points*
- *describe process of transmission and absorption, and reflection, refraction, diffraction and polarisation of waves*
- *distinguish clearly between longitudinal and transverse waves*
- *use the equation $v = f\lambda$ in calculations and describe spectra of electromagnetic radiation and sound*
- *demonstrate how an oscilloscope can be used to measure period and frequency*
- *explain that 'inverse square law' behaviour is a result of spreading of power over increasing area*
- *describe the law of reflection*
- *interpret refraction using rays and wavefronts*
- *describe how the diffraction of a wave at a gap depends on the wavelength and the size of the gap*
- *explain the meaning of phase and phase difference*
- *understand the health hazards presented by different bands of UV radiation*
- *explain the Doppler effect*

LEARNING SUMMARY

Source, journeys and interactions

AQA A	2	Edexcel context	2
AQA B	1	OCR A	2
CCEA	2	OCR B	2
Edexcel concept	2	WJEC	2

Mechanical waves, like sound waves, originate from sources that physically vibrate and the travel of the sound needs a medium in which there are particles to vibrate. For light waves, no physical material is involved, and the vibrations involve electric and magnetic fields.

Some kinds of absorption have a particularly significant impact. They change the material which they reach, as happens in photographic film, in the CCDs of digital cameras, and in human eyes. The absorption gives rise to detection.

Emission involves transfer of energy away from a source. Waves can transfer energy to material that stops, or absorbs, them. **Absorption** is the end of a journey for the energy, and so for the waves themselves. A defining feature of waves is that, although they carry energy, and can carry information in the process, no material travels from source to absorber.

The nature of the information carried by a wave depends on the nature of the source, and also on what happens during the journey. It is in this way, by light waves and sound waves, that we know of the world around us.

Different processes can happen on journeys of waves before they are absorbed. They may simply be **transmitted**, continuing on their journey with no change. Boundaries between materials may also **reflect**, **refract** or **diffract** waves. Within material, light can experience **polarisation** and **scattering**. Absorption itself also happens in materials, and it can happen either relatively abruptly soon after the waves cross boundaries into a material, or it can happen gradually.

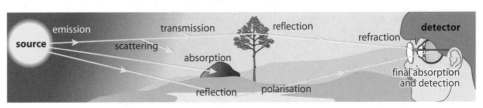

Waves, imaging and information

Wave motion

AQA A	2	Edexcel context	2
AQA B	1	OCR A	2
CCEA	2	OCR B	2
Edexcel concept	2	WJEC	2

The waves on a water surface are almost transverse, the particles move in an elliptical path.

Waves are used to transfer a signal or energy. They do this without an accompanying flow of material, although some waves, such as sound, can only be transmitted by the particles of a substance.

Sound and other compression waves are classified as **longitudinal** because the vibrations of the particles carrying the wave are along, or parallel to, the direction of wave travel. All electromagnetic waves are **transverse**; the vibrations of the electric and magnetic fields in these waves are at right angles to the direction of wave travel.

The diagram shows a longitudinal wave being transmitted by a spring and a transverse wave on a rope.

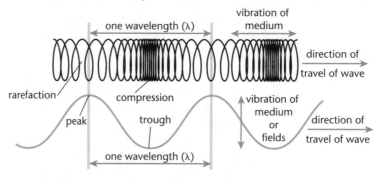

Wave measurements

AQA A	2	Edexcel context	2
AQA B	1	OCR A	2
CCEA	2	OCR B	1, 2
Edexcel concept	2	WJEC	2

These measurements apply to all **progressive** waves. A progressive wave is one that has a profile that moves through space.

- **Wavelength** (symbol λ) is the length of one complete cycle – see diagram above.
- **Amplitude** (symbol a) is the maximum displacement from the mean position, see diagram on page 92.
- **Frequency** (symbol f) is the number of vibrations per second, measured in hertz (Hz).
- **Speed** (symbol v) is the speed at which the profile moves through space.
- **Period** (symbol T) is the time taken for one vibration to occur. It is related to frequency by the equation $T = 1/f$.

> **KEY POINT**
>
> For all waves, the wavelength, speed and frequency are related by the equation:
>
> $$\text{speed} = \text{frequency} \times \text{wavelength}$$
> $$v = f \times \lambda$$

Representing waves

AQA A	2	Edexcel context	2
AQA B	1	OCR A	2
CCEA	2	OCR B	1, 2
Edexcel concept	2	WJEC	2

The diagram above shows representations of **longitudinal** and **transverse** waves. There are other ways of representing waves and their motion. The diagram here shows waves, which could be longitudinal or transverse, spreading out from a source.

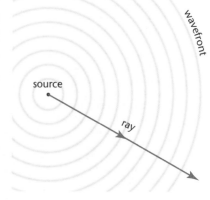

The **wavefront** can be thought of as the line of a crest or trough of a transverse wave or as a compression or rarefaction of a longitudinal wave. A **ray** shows the direction of travel. Note that as the waves spread from the source shown in the diagram on page 88, the rays 'radiate'. They behave rather like 'radii'.

Diagram **a**, below, shows a bird floating on water. The horizontal direction on the page corresponds to horizontal distance across the water. The vertical direction corresponds to vertical displacement. Diagram **b** shows a displacement–time graph for the motion of the bird. The basic shape of the curve, called a **sinusoidal curve**, is the same in both parts, but the way that information is represented is very different.

a

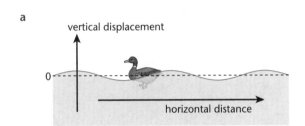

vertical displacement

0

horizontal distance

b

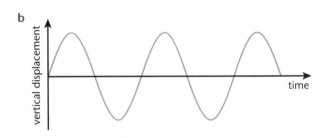

vertical displacement

time

Wave representations on a cathode ray oscilloscope

A common error is to misinterpret a CRO display as a graph of displacement against distance. This leads to a horizontal measurement, which represents time, being interpreted as a distance.

A **cathode ray oscilloscope** (CRO) can be used to display a representation of wave motion. The CRO plots the displacement of one point on the wave, such as the location of the bird in the diagram above, against time.

To measure the **period** and **frequency** of a wave from a CRO display, the time–base control, which controls the rate at which the dot sweeps the screen horizontally, needs to be in the 'cal' (calibration) position. The diagram below shows a CRO display of a sound wave, with the time–base set to $2 \, \text{ms} \, \text{cm}^{-1}$.

One cycle of the wave occupies a distance of 8 cm on the screen, so the period, $T = 8 \times 2 \, \text{ms} = 16 \, \text{ms} = 1.6 \times 10^{-2} \, \text{s}$.

The frequency of the wave, $f = 1/T = 1 \div (1.6 \times 10^{-2}) \, \text{s} = 62.5 \, \text{Hz}$.

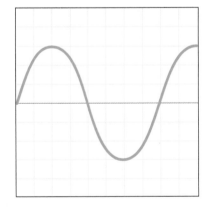

Intensity and the inverse square law

A point source such as a star radiates energy in all directions. In this case the variation of intensity with distance follows an inverse square pattern, with:

$$I \propto 1/r^2$$

$$I = P_0/4\pi r^2$$

where P_0 is the power of emission of a localised source.

The signal strength of a radio broadcast decreases the further away the receiver is from the transmitting aerial. In a similar way, a sound appears fainter and a light seems dimmer further away from their origins. This is due to energy becoming spread over a wider area as the wave travels away from its source.

The **intensity** with which a wave is received depends on the power incident on the area of the detector.

> **KEY POINT**
>
> Intensity = power per unit area measured at right angles to the direction of travel.
>
> $$I = P/A$$
>
> Intensity is measured in $W \, m^{-2}$.

The **intensity** of a wave is related to its amplitude:

intensity, $I \propto$ amplitude^2

Radiation flux is an alternative name for intensity.

The diagram opposite shows the sound spreading out from a loudspeaker. The ear that is further away detects a quieter sound because the power is spread over a wider area.

Note that **radiation flux** is a measure of the power at which radiation travels through unit area. It is also measured in W m^{-2}, and obeys the inverse square law.

Reflection and scattering

CCEA	2
Edexcel concept	2
Edexcel context	2
OCR A	2

We can use wavefronts or rays to represent the process of reflection at boundaries.

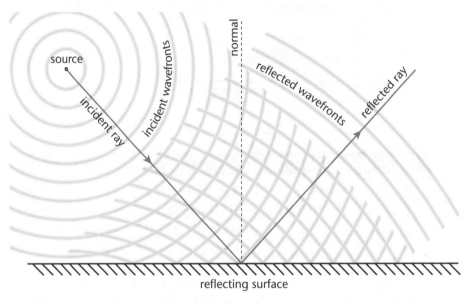

reflecting surface

Rays are particularly good for showing the law of reflection – the unfailing observation that mirrors reflect in a particularly regular way, so that **angle of incidence** and **angle of reflection** are the same. Note that the law of reflection applies to sound waves as well as to light waves.

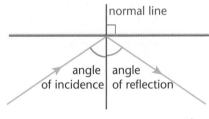

Surfaces that do not have the smoothness of mirrors do not reflect waves in a regular way. The rays become scattered. Scattering also occurs when waves experience reflection by many small particles that are suspended in the medium through which they are travelling. This is why you cannot see through mists, for example.

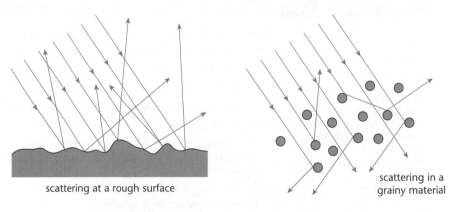

scattering at a rough surface

scattering in a grainy material

Refraction

AQA A	2	OCR A	2
CCEA	2	OCR B	1
Edexcel concept	2	WJEC	2
Edexcel context	2		

Refraction, like reflection, takes place at boundaries between materials. Again, diagrams with wavefronts and rays help us to visualise the process, but rays are better for an analysis that can incorporate actual values of angles. You can read more about refraction in section 3.3.

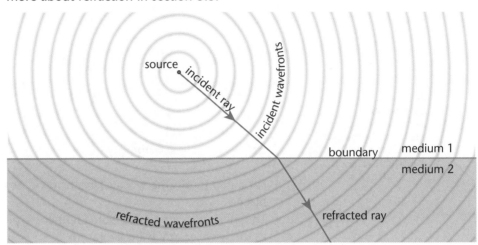

Polarisation

AQA A	2	Edexcel context	2
AQA B	1	OCR A	2
CCEA	2	OCR B	1
Edexcel concept	2	WJEC	2

Polarising material is used in some types of sunglasses and in filters for camera lenses. It cuts out reflected light from a flat surface such as an expanse of water.

Polarisation is a phenomenon that affects transverse waves only, and not longitudinal waves. Light waves are transverse, of course, and they are not normally polarised when emitted. In an unpolarised wave the vibrations are in all planes at right angles to the direction of travel.

The diagram opposite shows the vibrations in unpolarised and polarised waves. Light waves can become polarised as they pass through some materials and they are partially polarised when reflected.

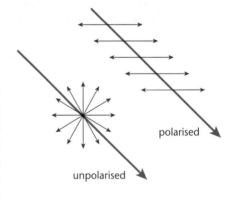

In order to receive a television or radio broadcast, the receiving aerial has to be lined up with that of the transmitter. This is because radio waves broadcast from aerials are polarised: the vibrations are only in one plane. For radio and television broadcasts the plane of polarisation is usually either vertical or horizontal.

When polarised light passes through a polarising material, its plane of polarisation is rotated, and some absorption takes place. Malus' law of polarisation provides a rule for predicting the intensity of transmitted light:

$$I = I_o \cos^2\theta$$

where I_o is the initial intensity and θ is the angle between the plane of polarisation of the initial light and the polarising axis of the material.

Diffraction

AQA A	2	Edexcel context	2
AQA B	1	OCR A	2
CCEA	2	OCR B	2
Edexcel concept	2	WJEC	2

The spreading out of waves as they pass through openings or the edges of obstacles is called diffraction. All waves can be diffracted.

The diagrams overleaf show what happens when waves pass through gaps of different sizes. The amount of spreading when the wave has passed through the gap depends on the relative sizes of the gap and the wavelength.

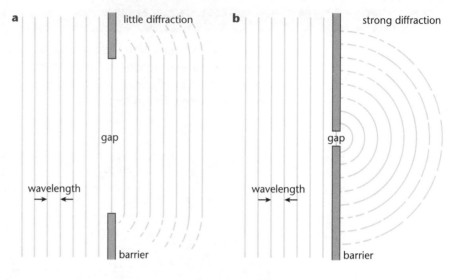

When answering questions about diffraction, always emphasise the size of the gap compared to the wavelength.

- For a gap that is many wavelengths wide, see diagram **a**, little spreading takes place.
- The maximum amount of spreading corresponds to a gap that is the same size as the wavelength as in diagram **b**.

Diffraction explains why you can 'hear' round corners but you cannot 'see' round corners.

Diagram **a** models what happens when light (wavelength approximately 5×10^{-7} m) passes through a doorway, which is much larger than the wavelength. Diagram **b** models sound (wavelength approximately 1 m) passing through the same doorway. There is much more diffraction when wavelength and gap are similar in size.

This also explains why long wavelength radio broadcasts can be detected in the shadows of hills and buildings, but short wavelength broadcasts cannot.

Phase and phase difference

AQA A	2	Edexcel context	2
AQA B	1	OCR A	2
CCEA	2	OCR B	2
Edexcel concept	2	WJEC	2

Both transverse and longitudinal waves can be represented by graphs of displacement against position. The position along a wave pattern can be measured in two ways:

- as a distance from a point in space
- as an angle.

In angular measure, one complete cycle of the wave is represented by an angle of 360°. This is shown in the diagram below.

A displacement–position or displacement–time graph does not distinguish between a transverse wave and a longitudinal one.

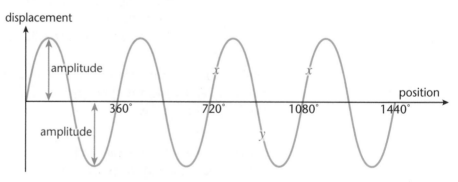

A displacement–position graph is like a slice through the wave at one moment, like a snapshot. At first sight it looks like a displacement–time graph, but it says very little about time. It is easy to confuse the two kinds of graph, but they provide very different information.

Using angular measure is a convenient way of comparing the **phase** of different parts of a wave. Two points on a wave are **in phase** if they have the same displacement and velocity. Points marked x on the diagram are in phase.

The point marked y is exactly **out of phase** with the points x as it has the opposite displacement and its velocity has the same value but in the opposite direction.

The **phase difference** between two points on a wave describes their relative displacement and velocity and is normally expressed in degrees or wavelengths. The point y is 180° or $\lambda/2$ out of phase with the points x.

Sound spectra

The **hertz**, or Hz for short, is the SI unit of frequency. One hertz is one complete 'event' or one cycle per second. $1\,Hz = 1\,s^{-1}$.

Take care – the spectrum of a particular sound is not the same as the complete spectrum of all possible sound frequencies.

The word **spectrum** means a range. The spectrum of human hearing, the audible range, is from about 20 Hz to 20 kHz. Other animals have different and often much wider ranges. Some dolphins and whales, for example, have audible ranges from a few hertz up to more than 200 kHz. Sound with frequency that is above the upper limit of the human range, above 20 kHz, is called **ultrasound**.

A pure note has a single frequency. Western musical notation divides sounds into octaves, with the frequency of the start of one octave being twice that of the start of the previous one.

Many sounds are not pure notes, but can be analysed as combinations of several frequencies. The frequency combination of a particular sound is sometimes called the 'spectrum' of that particular sound.

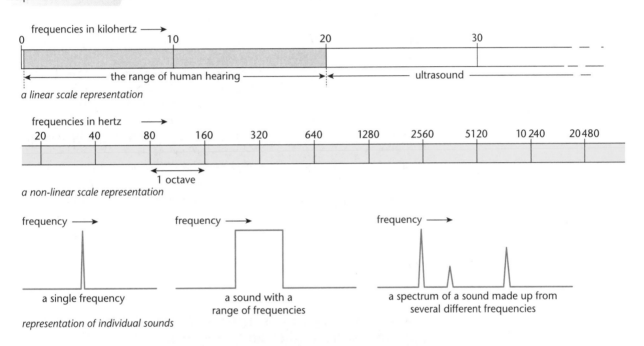

a linear scale representation

a non-linear scale representation

representation of individual sounds

The light spectrum

The speed of all electromagnetic waves in a vacuum is 2.999×10^8 m s⁻¹. The value 3.00×10^8 m s⁻¹ is normally used for both a vacuum and air.

The different changes of speed when light consisting of a range of wavelengths is refracted is responsible for dispersion, the splitting of light into a spectrum.

Visible light is a small part of a very wide spectrum. The electromagnetic spectrum consists of a whole family of waves, with some similarities and some differences in their behaviour.

Similar properties include:

- they consist of electric and magnetic fields oscillating at right angles to each other
- they travel at the same speed in a vacuum
- they are transverse waves, and they can be polarised
- they show the same pattern of behaviour in reflection, refraction, interference and diffraction.

The differences include:

- the shorter wavelength waves undergo a greater change of speed when being refracted
- diffraction is easier to observe in the behaviour of the longer wavelength waves
- interactions of light with matter, transmission (which is really an absence of interaction), absorption and reflection, vary considerably.

The diagram below shows the range of wavelengths and frequencies of the waves that make up the spectrum.

Microwaves are given here as a separate part of the spectrum, although they can be considered as short-wavelength radio waves.

frequency / Hz	10^{20}	10^{17}	10^{14}	10^{11}	10^8	10^5
	gamma rays	ultraviolet	infra-red		radio waves	
	X-rays		light	microwaves		
wavelength / m	10^{-12}	10^{-9}	10^{-6}	10^{-3}	1	10^3

The frequency and wavelength ranges of X-rays and gamma rays overlap. The difference between them is in the origin of the waves: an X-ray is emitted when a high-speed electron is suddenly brought to rest and a gamma ray is emitted from a nucleus, usually along with alpha or beta emission.

The table below shows typical wavelengths, the origins of the waves and the uses of the main parts of the electromagnetic spectrum.

Excited here means having energy to lose.

All objects give out infra-red radiation, no matter what the temperature. The hotter the object, the more power is emitted and the greater the range of wavelengths.

Name of radiation	Typical wavelength/m	How produced	Used for
gamma	1×10^{-12}	an excited nucleus releasing energy in radioactive decay	tracing the flow of fluids and treating cancer
X-rays	1×10^{-10}	high-speed electrons being stopped by a target	seeing inside the body
ultraviolet	1×10^{-8}	very hot objects and passing electricity through gases	security marking
light	1×10^{-6}	hot objects and electrically excited gases	vision and photography
infra-red	1×10^{-5}	warm and hot objects	heating and cooking
microwave	1×10^{-1}	microwave diode or oscillating electrons in an aerial	communications and cooking
radio	1×10^2	oscillating electrons in an aerial	communications

Transmission and absorption of ultraviolet, or UV, radiation by atmosphere, skin and sunscreen

UVc radiation from the Sun (wavelength approximately 100 to 280 nanometres) is mostly absorbed by the atmosphere. UVb (wavelength 280 to 320 nanometres) is partly but not completely absorbed by the atmosphere. It is then absorbed by the outer layers of skin, and this localised absorption means that it transfers energy rapidly to the skin. This can cause chemical damage, with short-term burning of the skin and increased risk of skin cancer in the longer term. UVa (wavelength 320 to 400 nanometres) is transmitted by the atmosphere, and is absorbed more gradually than UVb by the skin. So although it travels further through the skin and is still a cause of problems, it is less dangerous than UVb.

Commercial sunscreens are designed to absorb UVb, and the 'factor' provided by the supplier relates to UVb absorption and not to UVa. Some 'broad spectrum' sunscreens are available, however, and these are better absorbers of UVa.

Filters

AQA B   1
OCR A   2

Filters transmit some frequencies but allow others to pass through. Sunscreen filters UVb out of a stream of radiation. A red filter transmits low frequency visible light but absorbs higher frequencies. It makes the world seem rose-tinted.

It is more difficult to filter sounds, but electrical signals that represent sounds, or audio signals, can be filtered by electrical circuitry. Low pass filters allow low frequency audio signals to pass, but absorb higher frequencies. The signal from a low pass filter can be fed to a low frequency loudspeaker, or woofer, so that it does not vibrate efficiently at high frequencies. Likewise, a high pass filter transmits higher frequencies that can be fed to a high frequency loudspeaker, or tweeter, so that it can work most efficiently.

The Doppler effect

For all kinds of waves, relative movement of the source and an observer produces change in the frequency (and therefore wavelength as well) that the observer detects. This change is called the Doppler effect.

Doppler effect – moving observer:

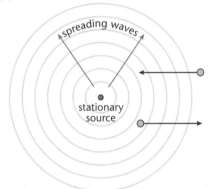

An observer moving towards a source passes through wavefronts more often than a stationary observer, and so measures a higher frequency.

An observer moving away from a source passes through wavefronts less often than a stationary observer, and measures a lower frequency.

Doppler effect – moving source:

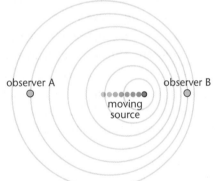

Each circular wavefront is centred on the location of the source at the moment that it was emitted.

Observer B detects a frequency that is higher than the frequency of emission.

Observer A detects a frequency that is lower than the frequency of emission.

Progress check

1 Describe the difference between a longitudinal and a transverse wave and give one example of each.

2 Calculate the frequency of a VHF radio broadcast that has a wavelength of 2.75 m. The speed of radio waves, $c = 3.00 \times 10^8 \, \text{m s}^{-1}$.

3 Two points on a wavefront have a phase difference of 180°. Describe their relative displacement and velocity.

3 The points have the same amount of displacement but in opposite directions, i.e. the displacement of each one is the negative of the displacement of the other one.
Similarly, the velocities are equal in size but opposite in direction.

2 1.09×10^8 Hz.

1 In a longitudinal wave the vibrations are parallel to the direction of travel, e.g. sound or any other compression wave.
In a transverse wave the vibrations are at right angles to the direction of travel, e.g. light or any other electromagnetic wave.

3.2 Sound and superposition

After studying this section you should be able to:

- *explain superposition*
- *describe how stationary waves are created on strings*
- *explain the harmonics of a string, and how they are related to the string's length and the sound's wavelength*
- *explain how beats are created*
- *explain how wave superposition causes interference patterns and describe the conditions needed for patterns to be observable*
- *compare the decibel scale with the range of audible intensities of sound*
- *describe sonar and ultrasound scanning techniques*

LEARNING SUMMARY

Superposition and stationary waves

AQA A	2	Edexcel context	2
AQA B	1	OCR B	2
CCEA	2	WJEC	2
Edexcel concept	2		

Sounds are made by vibrating sources, and most musical sounds are made by vibrating strings and pipes.

Simple mechanical waves can travel along ropes and along strings. A flick of a rope is enough to start a wave. If the wave reaches a fixed end of the rope or string, then its energy must be conserved. Some energy might pass out of the string, and be absorbed by a body that it is fixed to. Much of the energy can be reflected back along the spring. Such reflections can combine with the original wave to create **stationary or standing waves**. This process of combination, in which the total displacement of a point on the string is the sum of displacements caused by more than one wave, is called **superposition**. Superposition can take place with any type of wave, including light waves.

A stationary wave has a static profile. The waves on the vibrating strings and in the vibrating air columns of musical instruments are stationary waves.

Stationary waves are caused by the superposition of two waves of the same wavelength travelling in opposite directions. They often arise when a wave is reflected at a boundary, but the waves can come from two separate sources. The diagram shows the vibrations of a stationary wave on a string.

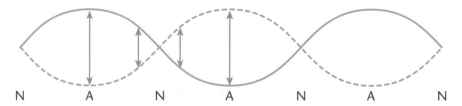

Stationary waves differ from progressive waves in a number of respects:

- there is no flow of energy along a stationary wave, although stationary waves often radiate energy
- within each loop of a stationary wave, all particles vibrate in phase with each other, and exactly out of phase (180° phase difference) with the particles in adjacent loops
- the amplitude of vibration varies with position in the loop
- there are **nodes** (points marked N in the diagram), where the displacement is always zero and **antinodes** (marked A in the diagram) which vibrate with the maximum amplitude.

A convenient way of measuring the wavelength of a progressive wave is to use it to set up a stationary wave, for example by superimposing the wave on its reflection from a barrier, and measuring the distance between a number of successive nodes or antinodes.

The wavelength of a stationary wave is twice the distance between two adjacent nodes or antinodes. The diagram opposite shows how a stationary wave (shown with a solid line) is formed from two progressive waves (shown with broken lines) travelling in opposite directions.

In **a** the waves' superposition takes place constructively, so each point on the wave has its maximum displacement.

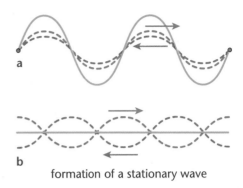

formation of a stationary wave

In **b** each progressive wave has moved one quarter of a wavelength; the superposition is now destructive, resulting (just for a moment) in no displacement at all points on the wave.

Harmonics

AQA A	2	Edexcel context	2
AQA B	1	OCR B	2
CCEA	2	WJEC	2
Edexcel concept	2		

A string that is fixed at its ends will have a node at each end. The simplest kind of stationary wave on a string has a node at each end and an antinode on the middle. It is called the string's **fundamental** note.

Along the length of the string, however, there could be one more node, or two more, or any number more. This means that the stationary wave can have different wavelengths, and different frequencies of vibration. These different ways of vibrating are called the **harmonics** of the string.

In practice, different harmonics can exist at the same time, to produce sounds that are more complex than simple notes. However, when a string is plucked, the higher frequency harmonics tend to fade most rapidly, so that the note changes, becoming more 'pure' as the single fundamental or first harmonic note becomes more dominant.

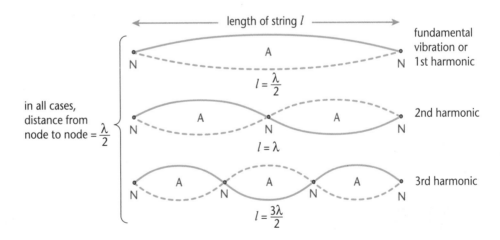

Factors affecting fundamental frequency of a string – Melde's apparatus

AQA B	1
Edexcel concept	2
Edexcel context	2

Melde's apparatus consists of a string or wire fixed at one end to a vibration generator, which provides a source of vibrations for which frequency can be continuously varied. The string passes over a pulley, and its other end carries a load, normally a set of slotted masses which can be used to vary the tension in the whole string. A small movable 'bridge' creates a fixed point for vibration of the string, so that the vibrating length of string can be measured.

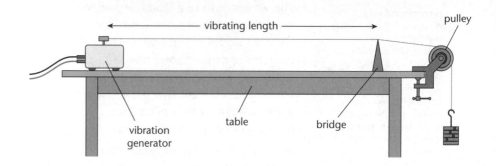

vibrating length — pulley — table — bridge — vibration generator

With the vibrating length fixed, the relationship between tension and fundamental frequency can be investigated. This provides a precise relationship in support of what every string musician knows: fundamental frequency, f, increases as tension, T, increases. In more detail:

$$f \propto \sqrt{T}$$

Similarly with fixed tension, the effect of length can be found:

fundamental frequency, f, decreases as length, l, increases, so that the proportionality is now inverse:

$$f \propto 1/l$$

By using strings of different mass per unit length, μ, it is possible to find out that:

$$f \propto 1/\sqrt{\mu}$$

The frequency of a vibrating string is related to its various physical circumstances by the equation:

$$f = \frac{1}{2l}\sqrt{\frac{T}{\mu}}$$

Superposition and beats

AQA B   1

Superposition of two notes of similar but not identical frequencies, f_1 and f_2, produces a fluctuation of sound intensity and loudness. The fluctuations are called **beats**.

The frequency of beat fluctuations is simply the difference between the frequencies of the two sources. If these frequencies become the same then the beats cease to exist. This point is used to tune instruments, using the instrument as one source of sound, f_1, and a tuning fork or other source of known and fixed frequency, f_2, as the other.

beat frequency = $f_1 - f_2$

When the beat frequency becomes zero the two frequencies, f_1 and f_2, are the same.

Superposition and interference

AQA A	2	Edexcel context	2
AQA B	1	OCR A	2
CCEA	2	OCR B	2
Edexcel concept	2	WJEC	2

Interference can occur with any type of wave, including water ripples and light waves.

Superposition of waves from two loudspeakers can produce a pattern of varying amplitudes of vibration called an **interference** pattern. Two identical sources produce remarkably neat patterns of alternating maxima and minima.

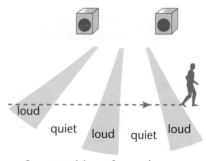

loud   quiet   loud   quiet   loud

Superposition of sound waves

The interference is **constructive** for transverse waves when troughs add together with other troughs, and crests coincide with crests, making larger troughs and larger crests. For longitudinal waves, similar addition of compressions and of rarefactions produces constructive interference. The interference is **destructive** when crests or compressions from one wave combine with troughs or rarefactions from the other to produce resulting vibration with minimised amplitude.

The diagram below represents the waves from two sources vibrating in phase. The solid and broken lines could represent peaks and troughs in the case of water waves or compressions and rarefactions in the case of sound waves.

> The amplitudes of waves from the two sources do not need to be the same, but they should be comparable for the interference to be observed. If one wave has a much greater amplitude than the other then the effects of cancellation and reinforcement will not be noticeable.

> Try drawing similar patterns and investigating the effect of changing the wavelength and the separation of the sources.

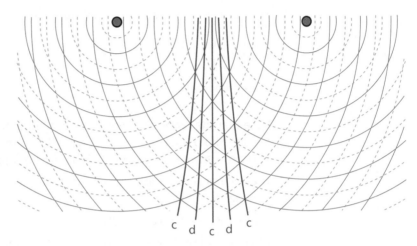

Constructive and destructive interference from two wave sources

Some lines of constructive interference **c** and destructive interference **d** have been marked in.

If the sources in the diagram are in phase, then at any point equidistant from both sources the waves arrive in phase, giving constructive interference. This is the case for points along the central **c** line in the diagram. At all other points in the interference pattern, there is a **path difference**, i.e. the wave from one source has travelled further than that from the other source.

> The conditions can be written as $n\lambda$ path difference for constructive interference and $(2n+1)\lambda/2$ path difference for destructive interference, where n is an integer 0, 1, 2, 3 ... and so on.

- **Constructive interference** takes place when the path difference is a whole number of wavelengths.
- **Destructive interference** takes place when the path difference is one and a half wavelengths, two and a half wavelengths, etc.

This is shown in the diagram.

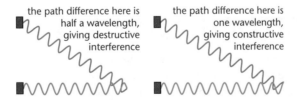

the path difference here is half a wavelength, giving destructive interference

the path difference here is one wavelength, giving constructive interference

These conditions for constructive and destructive interference only apply to two sources that are in phase. If the phase of one of the oscillators in the diagram at the top of this page is reversed so that the sources are in antiphase, then the conditions for constructive and destructive interference are also reversed. The effect is to shift the interference pattern so that lines of constructive interference **c** become destructive and vice versa. Any value of phase difference between the two sources gives an interference pattern that is stationary, provided that the phase difference is not changing. Two sources with a **fixed phase difference** are said to be **coherent**. Coherent sources must have the same wavelength and frequency.

Intensity, loudness and the decibel scale

AQA B   1
CCEA   2

Doubling the intensity (measured in Wm^{-2}) of a sound does not double its loudness. The loudness we hear is not proportional to the intensity. Apparent loudness is better (but still not perfectly) expressed in terms of **intensity level**, for which the unit is the **decibel**, dB.

On the decibel scale, the quietest sound that a person with normal hearing can perceive is 0 dB. It is sometimes called the threshold of hearing. On the decibel scale, an increase of a factor of 10 in the intensity adds 10 decibels, as in the table:

Intensity/ Wm^{-2}	10^{-12}	10^{-11}	10^{-10}	10^{-9}	10^{-8}	10^{-7}	10^{-6}	10^{-5}	10^{-4}	10^{-3}	10^{-2}	10^{-1}	10^{0}	10^{1}
Intensity level/dB	0	10	20	30	40	50	60	70	80	90	100	110	120	130

The threshold of hearing   The threshold of pain

Note that the range of intensities that the human ear can detect is very large, from the threshold of hearing at about $10^{-12}\,Wm^{-2}$ up to the threshold of pain at $10^{1}\,Wm^{-2}$. Instant perforation of the eardrum takes place at about $10^{4}\,Wm^{-2}$.

The relationship suggested in the table can be expressed more precisely as a formula:

$$\text{intensity level} = 10\log_{10} I/I_0$$

where I is intensity and I_0 is the intensity of the threshold of hearing

The situation is made more complicated by the fact that the threshold of hearing, and sensitivity of the ear in general, is different at different frequencies. Subjective judgement of loudness does not correlate perfectly with the objective decibel scale of sound intensity level. This is shown in **curves of constant loudness**.

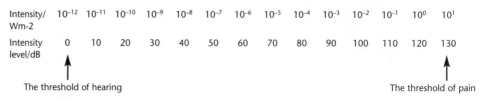

Sounds of different frequency with the same apparent loudness, shown by the two lines on the left, do not necessarily have the same intensity level, as shown in the curves on the right.

Sonar and ultrasound applications

CCEA   2

Bats find their way around by emitting pulses of sound and interpreting the reflected sounds. Sonar works in a similar way. The time between emission of a pulse and receiving reflection of the sound from a surface such as a sea bed provides a measure of the distance to the surface.

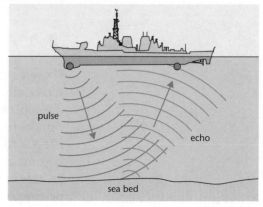

distance to surface = ½ time between emitting and receiving × speed of sound in the medium

The highest frequency of audible sound is about 20 kHz, which has a wavelength of 6 cm. That is similar in size to some larger body structures, and so significant diffraction will produce 'blurring' of images produced using audible sound waves like these. Medical ultrasound usually uses frequencies of several megahertz (MHz), which have wavelengths that are around 1 millimetre or less to prevent unwanted diffraction effects.

A sophisticated method of measuring the speed of an object takes advantage of the Doppler effect. Police speed checks, for example, can make use of radio waves, and the shift in frequency (the difference between emitted frequency which is already known and the reflected frequency that is detected by the apparatus) is proportional to the speed of the vehicle:

$$\Delta f \propto v$$

Ultrasound scans are used in medicine, particularly for examining unborn babies. They work on the same principle as sonar, but use higher frequencies of vibration. The pulses must be very short, since otherwise it would not be possible to distinguish between the emitted pulse and its reflection, and since the distances involved are much smaller than in ocean navigation and surveying.

Ultrasound scans use a single device or probe to emit pulses and to receive the echoes. In A scans, reflected peaks of different amplitude can be displayed. This can provide useful information about the nature of the reflecting boundaries within a patient's body.

A scan output

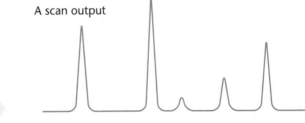

A scans are 'amplitude' scans.
B scans are 'brightness' scans.

B scan output

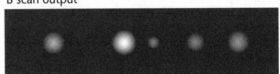

Progress check

1 When light passes through a slit which is 2mm wide, it produces a narrow beam. Explain why there is no observable diffraction.

2 For an observable interference pattern between two sources, the sources must be coherent.
  a What is meant by two coherent sources?
  b What TWO other conditions should be met for an observable interference pattern?

3 a The speed of sound in water is 1500 ms⁻¹. A ship's sonar system sends out a brief pulse of sound, and receives an echo from the sea bed 0.67 s later. What is the depth of the sea?
  b Suggest why sonar is less suitable for measuring long distances in air than in water.
  c Suggest why ultrasound cannot use audible frequencies.

3 a 500m
b Air is not a good medium for transmitting sound / the sound is absorbed over large distances / intensity of echoes in air is relatively low
c Pulses must be extremely short to avoid confusion of outgoing and returning pulse / with audible frequencies the duration of a pulse would be less than the period of vibration of the emitting scanner / a pulse cannot be produced without more than one vibration.
2 a Coherent means they have a fixed phase relationship.
b They should be the same type of wave and similar in amplitude.
1 The width of the slit is very large compared to the wavelength of light.

101

3.3 Light and imaging

After studying this section you should be able to:

- *use the relationships between refractive index, speeds of light, and angles of incidence and refraction*
- *describe how optical fibres are used in communication and medicine*
- *compare analogue and digital signals and how they deal with the problem of noise*
- *understand focal length and predict image type and location*
- *calculate magnifications*
- *perform calculations based on 'double-slit' interference of light*
- *explain the importance of diffraction of light to image resolution*
- *compare CT and MRI scanning technologies*

LEARNING SUMMARY

Refraction and refractive index

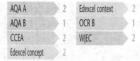

When waves cross a boundary between two materials there is a change of speed, called **refraction**. If the direction of wave travel is at any angle other than along the normal line, this change in speed causes a change in direction. The diagram opposite shows what happens to light passing through glass. Note that there is always some reflection at a boundary where refraction takes place.

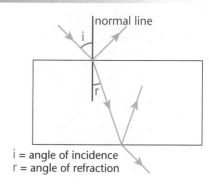

i = angle of incidence
r = angle of refraction

The change in speed is described by the **refractive index**.

> The refractive index of a boundary, $_1n_2$, is the ratio of the speed of light in the first medium, c_1, to the speed of light in the second medium, c_2.
> $$_1n_2 = c_1 / c_2$$
> Refractive index is a ratio, it does not have a unit.

KEY POINT

The **absolute refractive index** of a material is the factor by which the speed of light is reduced when it passes from a vacuum into the material. This is sometimes called the **absolute** refractive index (n) to distinguish it from the **relative** refractive index (n^{u2}) between two materials.

> The absolute refractive index of a material, n, is the ratio of the speed of light in a vacuum, c_v, to the speed of light in the material, c_m.
> $$n = c_v / c_m$$

KEY POINT

The greater the change in speed when light is refracted, the greater the change in direction. At a given angle of incidence, there is a greater change in direction when light passes into glass, with absolute refractive index $n_g = 1.50$, than when light passes into water, with absolute refractive index $n_w = 1.33$. The values quoted here are absolute values, but in practice there is little difference between the refractive index of a material relative to a vacuum and that relative to air.

The refractive index of air is 1.0003, so there is no difference in the refractive index of a material relative to air and relative to a vacuum when working to three significant figures.

When light is refracted:

- the incident light, refracted light and the normal all lie in the same plane
- **Snell's law** relates the change in direction to the change in speed that takes place.

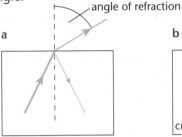

> Snell's law states that:
>
> $$\frac{\sin i}{\sin r} = \frac{c_1}{c_2} = {_1}n_2$$
>
> **KEY POINT**

So the sines of the angles are in the same ratio as the speeds in the media when refraction takes place.

Total internal reflection and critical angle

When light speeds up as it crosses a boundary, the change in direction is away from the normal line. For a particular angle of incidence, called the **critical angle**, the angle of refraction is 90°.

The diagrams below shows what happens to light meeting a glass–air boundary at angles of incidence that are **a** less than, **b** equal to, and **c** greater than, the critical angle.

> The intensity of the internal reflection increases as the angle of incidence increases.

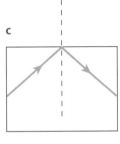

there is a weak reflection   there is a stronger reflection   all light is reflected

These diagrams show that:

- at angles of incidence less than the critical angle, both reflection and refraction take place
- at the critical angle, the angle of refraction is 90°
- at angles of incidence greater than the critical angle, the light is **totally internally reflected**.

> The abbreviation TIR is often used for total internal reflection.

Snell's law can be used to show that the relationship between the critical angle and refractive index is:

> A diamond has a very high refractive index and a low critical angle. Multiple internal reflections make diamonds sparkle.

> $$n = 1/\sin c$$
> where c is the critical angle and n is the absolute refractive index.
>
> **KEY POINT**

Fibre optics

Total internal reflection is used in prismatic binoculars, car and cycle reflectors and cats' eyes, but its greatest impact has been in the fields of medicine and communications.

Optical fibres enable light to travel round curves by repeated total internal reflection at the boundaries of the fibre, which is made from glass or plastic. Provided that light hits the boundary at an angle greater than the critical angle, none passes out of the fibre. The diagram opposite shows how light can be made to travel 'round the bend' of a fibre.

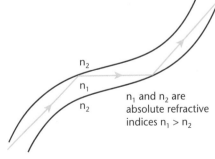

n_1 and n_2 are absolute refractive indices $n_1 > n_2$

In medicine, optical fibres are used in **endoscopy** to look inside a patient. This can be done using a natural body opening or by 'keyhole surgery', which requires a small incision just wide enough to take the fibre. Two bundles of fibres are used in endoscopy: one to transmit light into the patient and the other to carry the reflected light to a small television camera.

Although they are commonly referred to as light, the pulses used are usually in the infra-red region of the electromagnetic spectrum.

In communications, optical fibres are used to transmit telephone conversations, television and radio programmes in the form of **digital signals**. The signals used can only have the values 1 or 0, where 1 is represented by a pulse of light from a laser diode and 0 is the absence of light. The diagram opposite shows how optical fibres are used in telephone communications.

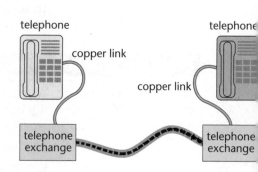

The range of a signal in a copper cable is only a few km, compared to up to 100 km in a glass fibre.

The advantages of using optical fibres rather than copper cables in communications are:

*   the range of a signal (the distance it can travel) in an optical fibre is much greater than that in a copper cable, so amplification is needed less frequently
*   optical fibres do not pick up **noise** from changing magnetic fields.

However, the signal in an optical fibre is still subject to distortion as it travels along the fibre. This is due to **multimode dispersion**, a process which causes the pulse to become elongated as some parts of the pulse travel further than others, for example when the cable goes round corners. In modern fibres the effect of this is reduced by concentrating the light pulses along a very narrow core of fibre, so that there is only one effective route for the light. These are **monomode** fibres.

A similar effect occurs if 'light' containing more than one wavelength is used, as the different wavelengths travel at different speeds in the fibre.

Multimode dispersion and its effect on a digital signal

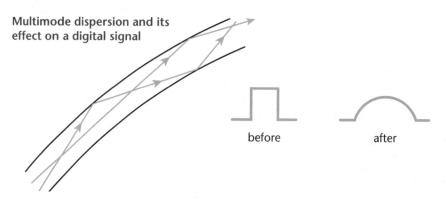

Noise and analogue and digital signals

AQA B   1
OCR B   1

Noise or distortion can be easily removed from digital signals in a process called regeneration.

Noise affects the amplitude of a signal. In AM (amplitude modulated) transmissions, the information is carried in the form of variations in amplitude of a wave. Any noise cannot be distinguished from the signal and so cannot be removed. When the signal is amplified, the noise is also amplified.

This is not the case with a digital signal, which only has certain allowed values. If the amplitude varies around these values, it can be restored. This is shown in the diagram on the next page.

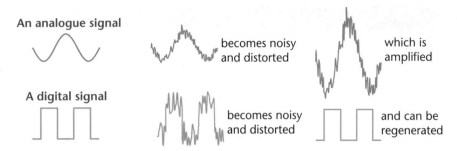

The 'allowed values' used in communications are normally 1 and 0, so regeneration is a straightforward task.

Converging lenses, objects and images

Refraction takes place at the surfaces of lenses. This can be shown using wavefront diagrams or ray diagrams:

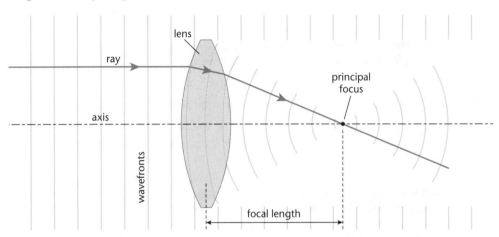

Both types of diagram show that a narrow parallel beam of light arriving at a lens is brought to a focus, or converged, near a single point. This point is called the **principal focus** of the lens. The distance from the centre of the lens to its principal focus is called its **focal length**.

Note also that a small source of light placed at the principal focus of a converging lens can produce a parallel beam of light from the lens.

Ray diagrams are particularly useful for analysing and predicting the nature of images that will be produced by different lenses.

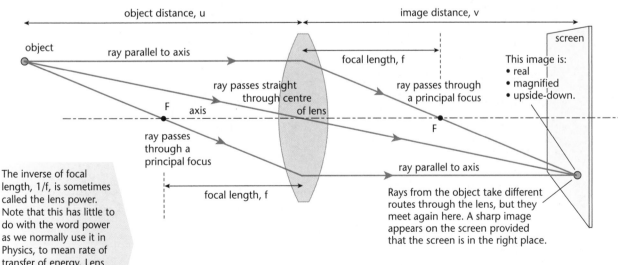

The inverse of focal length, 1/f, is sometimes called the lens power. Note that this has little to do with the word power as we normally use it in Physics, to mean rate of transfer of energy. Lens power has units m^{-1}, and $1\,m^{-1}$ is given the name of 1 dioptre in this context.

Relative distances of object and image from the lens can also be calculated, provided that the focal length of the lens is known, using the formula:

$$1/f = 1/v + 1/u$$

Different kinds of image

CCEA	2
OCR B	1

A lens can produce an image that is an 'upside-down' or inverted view of the object. The object might also be magnified, or diminished. **Linear magnification** is the ratio of the size of an image to the size of an object. As the diagram on the previous page suggests, this is equal to the ratio of image distance, v, to object distance, u.

linear magnification, m = image size / object size = v / u

Real images can be projected onto screens, as happens in an eye, a camera or a cinema. **Virtual** images cannot be projected onto screens but must be viewed directly. In order to trace the location of a virtual image in a ray diagram, lines must be drawn that are not real rays but projects of rays. An image in an ordinary mirror, for example, is a virtual image, as is the image produced by a converging lens when it is used as a magnifying glass.

Linear magnification

CCEA	2
OCR B	1

Linear magnification provides a way of comparing image size with object size.

linear magnification, m = image size ÷ object size

The ratio of image size to object size is the same as the ratio of image distance, v, to object distance, u. So we can also write:

$$m = v/u$$

Double-slit interference

AQA A	2	OCR B	2
AQA B	1	WJEC	2
CCEA	2		
OCR A	2		

Diffraction of light is, as for any other kind of wave, fundamentally a quite simple process. Waves spread out when passing through a gap or passing an obstacle. However, different parts of a beam of light act as different sources, so that interference takes place. This happens both at gaps and at small obstacles.

A colour filter can be used between the lamp and the slits to reduce the range of wavelengths interfering, but this also reduces the intensity of the interference pattern.

Interference patterns are easy to set up with water waves and sound because two sources can be driven from the same oscillator, giving coherence. Because of the way in which light is emitted in random bursts of energy from a source, it is not possible to have two separate sources that are coherent. Instead, diffraction is used to obtain two identical copies of the light from a single source. This is done by illuminating two narrow slits with a lamp that is parallel to the slits so that the same wavefronts arrive at each slit. A suitable arrangement is shown in the diagram below.

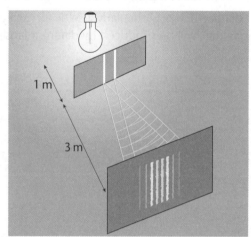

Bright and dark fringes on the screen are due to constructive and destructive interference between the two overlapping beams of light. The separation of the fringes depends on:

- the separation of the slits
- the wavelength of the light
- the distance between the slits and the screen.

This formula enables the wavelength of light to be measured from a simple experiment. The fringe spacing, x, should be obtained by measuring the separation of as many fringes as are visible and dividing by the number of fringes.

The separation of the fringes is related to the other variables by the formula:

$$\text{fringe spacing} = \frac{\text{wavelength} \times \text{distance from slits to screen}}{\text{slit separation}}$$

$$x = \lambda D/a$$

where x is the distance between adjacent bright (or dark) fringes
λ is the wavelength of the light
D is the distance between the slits and the screen
a is the distance between the slits.

As the wavelength of light is very small, the slits need to be close together and separated from the screen by a large distance for the interference fringes to be seen.

Diffraction gratings

AQA A	2	OCR B	2
AQA B	1	WJEC	2
CCEA	2		
OCR A	2		

A diffraction grating is a piece of material, such as a glass slide, with parallel lines engraved very close together onto its surface. Light passing through such a grid or grating of lines is experiencing many gaps and obstacles. Diffraction takes place, so that the grating acts as an array of sources of light. Interference then takes place between light from the gaps, producing patterns of maximum and minimum intensity. The close spacing of the gaps results in large spacing of the maxima and minima.

The relationship between the spacing of the gaps and the angles at which maxima appear is:

$$n\lambda = d \sin\theta$$

λ is the wavelength of the light
n = 0 for the central maximum, 1 for the next maximum, and so on
d is the grating separation
θ represents the angles at which the maxima appear

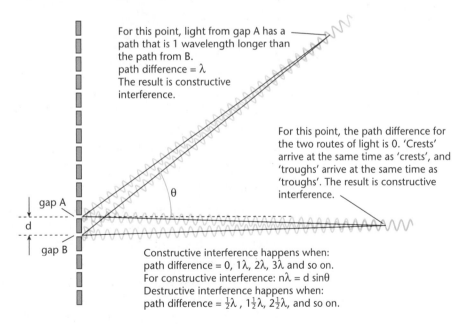

For this point, light from gap A has a path that is 1 wavelength longer than the path from B.
path difference = λ
The result is constructive interference.

For this point, the path difference for the two routes of light is 0. 'Crests' arrive at the same time as 'crests', and 'troughs' arrive at the same time as 'troughs'. The result is constructive interference.

gap A

d

gap B

Constructive interference happens when:
path difference = 0, 1λ, 2λ, 3λ and so on.
For constructive interference: $n\lambda = d \sin\theta$
Destructive interference happens when:
path difference = $\frac{1}{2}\lambda$, $1\frac{1}{2}\lambda$, $2\frac{1}{2}\lambda$, and so on.

Note that the appearance of wavelength in the equation tells us that different colours of light produce maxima and minima at different angles. Thus diffraction can disperse white light into a spectrum. This is the source of the colours seen when looking at a CD surface – the surface produces diffraction by reflection rather than transmission in this case, but the principle is the same.

Light of a single wavelength or frequency has a single colour, and is called **monochromatic**.

Image resolution

OCR B   1

Diffraction effects produce blurring. Smaller light collecting lenses produce greater diffraction effects. The lenses are said to have small **apertures**. Large aperture lenses produce sharper images that are better at distinguishing two objects that are close together – they are better at **resolving**.

CT and MRI scanning

CCEA   2

X-ray imaging of the human body has been possible for more than 100 years. Simple 'shadow' images rely on stronger scattering and absorption of X-rays by larger atoms, particularly the calcium atoms in bone. For other types of body tissue, differences in atom size and density are more subtle, but they do exist. CT scans take advantage of these differences.

CT stands for computer tomography. It requires a large number of separate X-ray exposures or scans, from different directions through a single slice of the body. Data from the scans is then combined to produce an image of the body slice. Dyes that accumulate in some tissues more than others and increase the scattering or absorption of the X-rays can make images clearer, and allow the process to take place without very large doses of radiation to the patient.

MRI stands for magnetic resonance imaging. Strong and varying magnetic field is used to induce vibration within atoms. As a result, the atoms emit radio waves. The nature of the vibration, and so the nature of the radio waves, depends on the kinds of atom involved. So different structures with different chemical compositions can be identified.

Progress check

The table gives the speed of light in different media.

Medium	Speed of light/$m\,s^{-1}$
vacuum	3.00×10^8
ice	2.29×10^8
ethanol	2.21×10^8
quartz	1.94×10^8

1 **a** Calculate the values of the absolute refractive indexes of ice, ethanol and quartz.
   b Calculate the refractive index for light travelling from ice into ethanol.

2 A piece of ice is placed in ethanol.
   a In which direction is light travelling for total internal reflection to take place?
   b Calculate the value of the critical angle.

3 The critical angle for a certain glass is 39°. Calculate the refractive index of the glass.

4 In a two-slit interference experiment using light, the slits are separated by a distance of 1.2 mm. The distance from the slits to the screen is 3.0 m and the separation of the bright fringes is 1.4 mm.
   Calculate the wavelength of the light.

5 A converging lens with focal length 15 cm is used to produce an image on a screen of an object that is 20 cm away from it.
   a Sketch a ray diagram and use it to describe the image.
   b Calculate the image distance.
   c Calculate the linear magnification.

c 3
b 60 cm
5 **a** Real, inverted, magnified
4 5.6×10^{-7} m
3 1.59
b 75°
2 **a** from ethanol to ice
b 1.04
1 **a** ice 1.31 ethanol 1.36 quartz 1.55

3.4 Information transfer

After studying this section you should be able to:

- distinguish between different ways of encoding and transmitting information
- explain the terms 'bandwidth', 'AM', 'FM', 'bit', 'sampling' and 'multiplexing'

LEARNING SUMMARY

Carrier waves, AM and FM

AQA B   1

Waves can transfer information as well as energy. A simple wave carries information about its frequency, wavelength and amplitude. It can carry much more information in the form of patterns of amplitude or patterns of frequency. The first type of pattern is amplitude modulation and the second is frequency modulation.

For amplitude modulation, or AM, the frequency (and wavelength) of the wave can be fixed. A radio set must be tuned in to that frequency when AM is used for radio transmission.

Frequency modulation radio transmission, or FM, can use a constant amplitude of wave, but needs to use a band or range of frequencies so that varying patterns of frequency can carry the encoded signal. The greater the band of frequencies, the more encoding possibilities there are and the faster information can be transmitted. This introduces the concept of bandwidth.

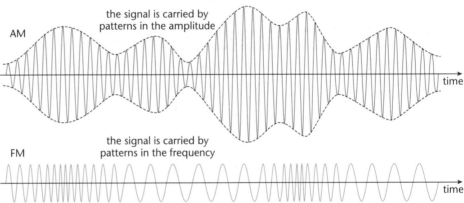

The frequencies here vary considerably, so this broadcasting is being done with a large bandwidth.

Analogue and digital signals

OCR B   1

Analogue measurement uses variation in one quantity to represent another. An ordinary clock uses the angle turned through by a hand of the clock to represent time. In amplitude modulation communication, the amplitude of the wave represents something else, such as voltage, which may in turn represent the changing pressure of air at a microphone. In general, analogue values can vary continuously, to represent the continuous variation in the original quantity such as voltage or air pressure.

The binary system of counting is simpler than the one we are familiar with – it is based on just two symbols, 0 and 1, unlike the familiar system that needs 10 symbols: 0, 1, 2, 3, 4, 5, 6, 7, 8, 9. Electrical systems can deal with just two possibilities much better than they can deal with 10 possibilities.

Digital signals do not vary continuously, but use just two values of a variable, such as light intensity or voltage. The two values are usually given the names 0 and 1. These are called binary digits, or bits.

A bit can have two values, so the information it can carry is very limited, but a series of bits can carry much more. A series of 8 bits is called a byte.

In one byte, there are 256 possible patterns of 0 and 1. 256 is 2^8.

In general:

number of possible different patterns = $2^{\text{number of bits}}$

A kilobyte is 1024 bytes. That is 1024×8 bits. $2^{1024 \times 8}$ is a very large number, and it can carry a lot of information.

Digital images, such as those produced by digital cameras, are broken into small squares or pixels. The information about each pixel can be stored as sets of numbers. One set of numbers can refer to the brightness of the pixel. The set of numbers could be 0 to 255, allowing 256 levels of brightness. Then one byte of information is enough to carry that information. Likewise, colours can be encoded as numbers.

Information storage

AQA B   1
OCR B   1

Information can be stored in analogue format or in digital format. Vinyl records, for example, use the shape of the groove in the plastic as a direct analogue representation of the sound.

CDs also use shape, having pits and spaces that reflect or do not reflect a laser beam. The detector either receives a reflected beam or it does not – it receives a binary digit 1 or a binary digit 0. A CD is a digital storage medium. DVDs are similar, but have more closely packed pits so that they can store more information.

Analogue to digital – sampling

AQA B   1
OCR B   1

An analogue signal can be turned into a digital signal by measuring at regular intervals, or sampling.

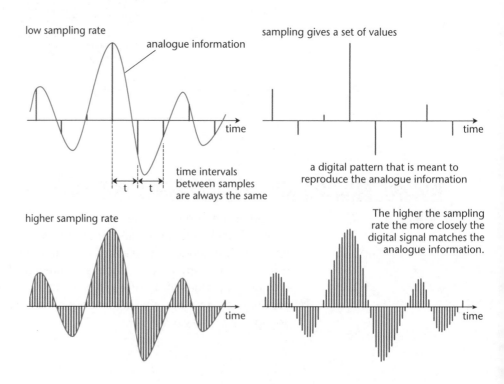

low sampling rate

analogue information

time

time intervals between samples are always the same

sampling gives a set of values

time

a digital pattern that is meant to reproduce the analogue information

higher sampling rate

time

The higher the sampling rate the more closely the digital signal matches the analogue information.

time

Transmission rate

AQA B   1
OCR B   1

Transmission rate or **bit rate**, measured in bits per second, provides a measure of speed of transfer of information. A film has sound and images, so the information needed for a movie display, by DVD or the internet, must be available more quickly than for listening to sound only. The rate of transmission must be higher for the DVD. Transmission rate can be measured in bits per second.

A CD player can take information from a disc at a rate of roughly 1 megabit s^{-1}. A Blu-ray player can transfer information at about 50 megabit s^{-1}, using a blue laser beam because its shorter wavelength allows it to read from very tightly spaced pits on the disc surface.

Time division multiplexing

AQA B   1

Multiplexing is the sending of many streams of information from different sources through a single system.

In **frequency division multiplexing**, the different streams use different small ranges of frequency of a wider band. The different streams travel at the same time.

In time division multiplexing, streams are broken up into very short sections, and each of these takes turns with sections from other streams.

Two streams of information…

… are separated into segments…

…and transmitted in turn through an information transfer medium (such as optical fibre)…

…at the receiving end the two sets of segments are separated.

Compression

AQA B   1

Computer memory space is limited, and so is the rate at which information can travel. Smaller amounts of information use less space, and travel more quickly. File compression is a process of reducing the amount of information from an original source. A principal way to do this is by looking for redundancy, which means needless repetition of information. In the text in this section, several words, such as 'information', appear more than once. Multiple appearances become redundant if the first one is given a code and later appearances use the same code.

Progress check

1 In what ways, when you speak, do you use,
   a amplitude modulation,
   b frequency modulation?

2 Many mobile phones take photographs that can then be transmitted to another phone. Explain the effect on this of:
   a Having high resolution images, with a large number of pixels per square centimetre.
   b Having images with a very large range of possible brightness values.

3 a Explain why a larger bandwidth allows a higher transmission rate.
   b Explain why a higher transmission rate allows a more faithful reproduction of information.

1 a To vary loudness b To vary pitch
2 a The more pixels per square centimetre, for a given light collecting area and for a given amount of information per pixel, the more information there is. This means that transmission rate must be high or time for sending and receiving will be long.
   b The larger the range of brightnesses the more bits must be used to send information about each pixel. Again, this requires high transmission rate or long time.
3 a Larger bandwidth allows more frequencies to be used for sending signals, so information can be sent more quickly.
   b Higher transmission rate allows a higher sampling rate, so that the digital signal more closely matches the original information.

Sample question and model answer

More students lose marks through poor communication than through poor maths. Use large sketches with clear labels as much as possible. Labels can be easier to write and sequence than long sentences. Examiners appreciate concise answers. Note that part (d) can be answered fully with a well-labelled sketch.

(a) What do ultrasound and ultraviolet light have in common when compared with audible sound and visible light, other than that we cannot detect them directly? [2]

(Ultrasound and ultraviolet light both have) higher,   1 mark
frequency (than the corresponding audible and visible radiations).   1 mark

(b) Explain why ultrasound, and not audible sound, is used for medical scanning. [5]

The wavelength of audible sound is long(er than for ultrasound),   1 mark
indication of how long, e.g. several centimetres or more.   1 mark
It is comparable to the size of the body and internal organs and   1 mark
diffraction effects are strong,   1 mark
producing blurring/reducing resolution.   1 mark

(c) What is the role of diffraction in the formation of an interference pattern by a 'diffraction grating'? [2]

A diffraction grating has many/alternating lines/gaps that transmit light.
Each line/gap is small/narrow enough for strong diffraction effect. Each
line/gap acts as a source of coherent light.   any 2 points, 1 mark each

(d) Explain how a diffraction grating produces bands of colour, with large angular separations that are dependent on wavelength. [5]

$n\lambda = d\ \sin\theta$ 1 mark

Significance of n as 0th, 1st, 2nd and so on order maxima, stated or
shown on sketch.   1 mark
Significance of d as line separation, stated or shown on sketch.   1 mark
Significance of θ as angular separation (from 0th order diffraction or
system axis), stated or shown on sketch.   1 mark
Small value of d (relative to wavelength) results in large value of θ
for any given value of n, stated or shown on sketch.   1 mark

Practice examination questions

1 (a) Distinguish between a **transverse** and a **longitudinal** wave and give one example of each. [4]

(b) Light can be **polarised** when it is reflected at a surface.
   (i) Describe the difference between polarised light and unpolarised light. [2]
   (ii) Explain why sound cannot be polarised. [1]

2 (a) State the laws that describe the refraction of light. [2]

(b) Light of frequency 6.00×10^{14} Hz is incident on an air–water boundary at an angle of 63°. The speed of light in air is 3.00×10^{8} m s^{-1} and the speed of light in water is 2.25×10^{8} m s^{-1}.
   (i) Calculate the wavelength of the light in air. [3]
   (ii) Calculate the refractive index for light passing from air into water. [2]
   (iii) Calculate the angle of refraction in the water. [2]

3 The diagram shows how the profile of a wave on a rope changes. In a period of 2.50×10^{-4} s it moves from position 1 to position 2.

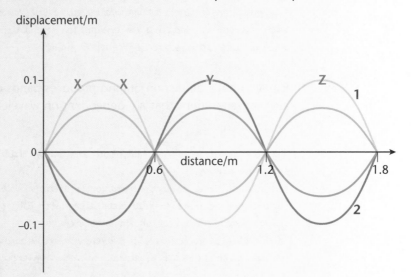

(a) (i) Write down the wavelength of the wave. [1]
   (ii) Calculate the frequency of the wave. [2]
   (iii) Calculate the speed of the wave along the rope. [3]

(b) (i) How can you tell from the diagram that the wave is a stationary wave? [1]
   (ii) Suggest how the stationary wave is formed in this example. [2]

(c) What is the phase difference between:
   (i) the two points marked X on the diagram? [1]
   (ii) the points marked Y and Z on the diagram? [1]

Give your answers in degrees.

4   The diagram shows two identical loudspeakers, A and B, placed 0.75 m apart. Each loudspeaker emits sound of frequency 2000 Hz.
Point C is on a line midway between the speakers and 5.0 m away from the line joining the speakers. A listener at C hears a maximum intensity of sound. If the listener then moves from C to E or D, the sound intensity heard decreases to a minimum. Further movement in the same direction results in the repeated increase and decrease in the sound intensity.

speed of sound in air = 330 m s^{-1}

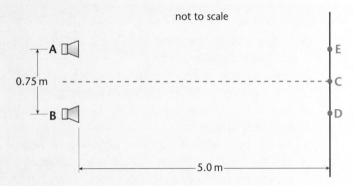

(a) Explain why the sound intensity is:
     (i)   a maximum at C
     (ii) a minimum at D or E.   [2]

(b) Calculate:
     (i)   the wavelength of the sound
     (ii) the distance CE.   [2]

5   The diagram shows an optical fibre made of a plastic-coated glass.
The refractive index of the glass is 1.58 and that of the plastic is 1.24.

(a) Calculate the refractive index for light striking the glass–plastic interface.   [2]

(b) Calculate the value of the critical angle at this interface.   [2]

(c) The diagram shows two paths of light through the fibre.
     Explain how this can affect a pulse of light as it passes along a fibre.   [2]

Chapter 4

Waves, particles and the Universe

The following topics are covered in this chapter:

- *The particle structure of matter*
- *Nuclei and radioactivity*
- *Light and matter*
- *The wider Universe*

4.1 The particle structure of matter

After studying this section you should be able to:

- *outline evidence for the existence of particles in matter, including the existence of atomic nuclei*
- *relate the four forces (gravitational force, electromagnetic force, strong and weak nuclear forces) to behaviour of matter*
- *explain that quarks, electrons and neutrinos are fundamental particles with antiparticles*
- *explain that hadrons are made of quarks, and name examples of hadrons*
- *describe the roles of exchange particles, including photons, gravitons, gluons, pions and intermediate vector bosons, in all forces between bodies*
- *use Feynman diagrams to represent particle exchange*

LEARNING SUMMARY

The existence of particles

AQA A	1, 2	Edexcel context	1, 2
AQA B	1, 2	OCR A	1, 2
CCEA	1, 2	OCR B	1, 2
Edexcel concept	1, 2	WJEC	1, 2

The kinetic model pictures a gas as being made up of large numbers of individual particles in constant motion. At normal pressures and temperatures the particles are widely spaced compared to their size. The motion of an individual particle is:

- **rapid** – a typical average speed at room temperature is $500\,\text{m}\,\text{s}^{-1}$
- **random** in both speed and direction – these are constantly changing due to the effects of collisions.

Gases exert **pressure** on the walls of their container and any other objects that they are in contact with. This pressure acts in all directions and is due to the forces as particles collide and rebound.

Brownian motion was first observed by Robert Brown while studying pollen grains suspended in water. The explanation of Brownian motion is due to Einstein.

Evidence for this movement of gas particles comes from **Brownian motion**; the random, lurching movement of comparatively massive particles such as smoke specks when suspended in air. This movement can only be attributed to the bombardment by much smaller air particles which are too small to be seen by an optical microscope and must therefore be moving very rapidly.

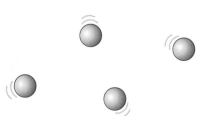

Gas particles are widely spaced. Their movement is random in both speed and direction.

Further evidence comes from diffusion – materials can mix because they do not have continuous structure, and all that is needed is mobility of their particles. The fixed and relatively simple ratios of masses of materials taking part in chemical equations is also explained by supposing that the materials have particle structures.

Particles can change their arrangements, resulting in different states or phases of matter.

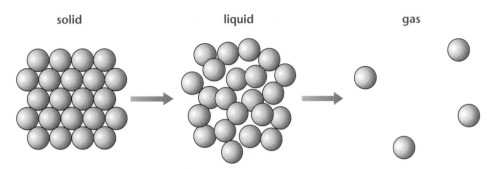

solid   liquid   gas

The diagrams above show particles that could be complex molecules or simple atoms. The particles are shown as spheres, for the sake of simplicity. But it is possible to explore inside atoms, to find out about their structure.

Nuclei and electrons

AQA A	1	Edexcel context	2
AQA B	1	OCR A	2
CCEA	2	OCR B	2
Edexcel concept	2	WJEC	2

Atoms are the building blocks of everyday material. In gold, for example, all of the atoms are essentially the same. About a century ago, people were trying to find out about the structures of atoms. They were trying to 'model' atoms.

The discovery of electrons, as part of matter, led to the idea of these being embedded in a spongy ball of positively charged matter. This idea was called the **plum pudding model** of the atom.

> An alpha particle is a positively charged particle consisting of two protons and two neutrons. The experiments were carried out in a vacuum so that the alpha particles were not scattered by air particles.

A later atomic model pictures the atom as a tiny, positively charged nucleus surrounded by negatively charged particles (electrons) in orbit. Evidence for this model comes from the scattering of alpha particles as they pass through a thin material such as gold foil. The results of these experiments, first carried out under the guidance of Ernest Rutherford in 1911, can be summarised as:

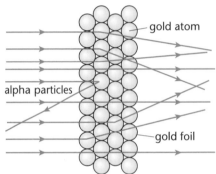

gold atom

alpha particles

gold foil

- most of the alpha particles travel straight through the foil with little or no deflection
- a small number are deflected by a large amount
- a tiny number are scattered backwards.

Rutherford concluded that this **alpha scattering** showed that the atoms of gold are mainly empty space, with tiny regions of concentrated charge. This charge must be the same sign as that of alpha particles (positive) for the back-scattering to be due to the repulsion between similar-charged objects.

There are two types of particle in the nucleus, **protons** and **neutrons**.

> You will not find El in the periodic table – it is fictitious.

> **KEY POINT**
> The nucleus of an element is represented as $_Z^A$El.
> Z is the atomic number, the number of protons.
> A is the mass number, the number of nucleons (protons and neutrons).

The element is fixed by Z, the number of protons. A neutral atom has equal numbers of protons in the nucleus and electrons in orbit. Different atoms of the same element can have different values of A due to having more or fewer neutrons. They are called **isotopes** of the element. As this does not affect the number of electrons in a neutral atom, the chemical properties of these atoms are the same. The most common form of carbon, for example, is carbon-12, $_6^{12}$C, which has six protons and six neutrons in the nucleus. Carbon-14, $_6^{14}$C, has the same number of protons but two extra neutrons. These two forms of carbon are isotopes of the same element.

The charges and relative masses of atomic particles are shown in the table below. The masses are in atomic mass units (u), where $1u = \frac{1}{12}$ the mass of a carbon-12 atom = 1.661×10^{-27} kg. The charges are relative to the value of the electronic charge, $e = 1.602 \times 10^{-19}$ C.

> The phrase 'relative to the value' means compared to the actual amount of charge, ignoring the sign.

Atomic particle	Mass	Charge
proton	1.01	+1
neutron	1.01	0
electron	5.49×10^{-4}	−1

> The ratio of charge to mass of a body, q/m, is called its specific charge. The specific charge of electrons is much larger than that of protons, because electrons have much less mass.

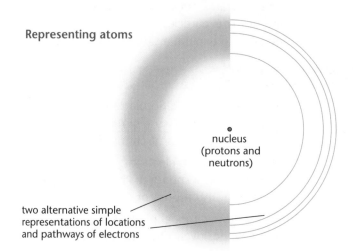

Representing atoms

nucleus (protons and neutrons)

two alternative simple representations of locations and pathways of electrons

Evidence from alpha-scattering experiments shows that the diameter of a nucleus is approximately one ten-thousandth (1×10^{-4}) that of the atom and the nuclear volume is 1×10^{-12} that of the atomic volume. As almost all the atomic mass is in the nucleus, it must be extremely dense.

Electron diffraction experiments allow more precise estimates of the size of nuclei to be made. The results of these experiments show that:

- all nuclei have approximately the same density, about 1×10^{17} kg m^{-3}
- the greater the number of nucleons, the larger the radius of the nucleus.

The four forces

AQA A	1
AQA B	1
WJEC	2

The forces that we experience every day can be classified as one of two types:

- **gravitational forces** affect all objects that have mass – they are very weak and have an infinite range
- **electromagnetic forces** also have an infinite range and are much stronger than gravitational forces – they include all forces due to static or moving charges.

> On the nuclear scale, gravitational forces are so weak that they can be ignored.

The nucleus is very concentrated in terms of both mass and charge. The electrostatic force between the protons is immense and the gravitational attraction is tiny, so if these were the only forces acting, the nucleus would be unstable. In addition to the forces above, there are two other fundamental forces.

> The strong force decreases very rapidly with increasing separation of the particles.

> The word 'particle' is now used for comparatively huge bodies such as molecules as well as 'fundamental' bodies that can't be split into anything smaller, such as electrons.

- The **strong nuclear force** affects protons and neutrons but not electrons. Its limited range, 1×10^{-15} m, means that it only acts between nucleons that are very close, i.e. next to each other in the nucleus.
- The **weak nuclear force** affects all particles. It has an even shorter range than the strong force, 1×10^{-18} m, and is responsible for beta decay.

It is the strong nuclear force that keeps the nucleus together. The strong attraction between the nucleons balances the electrostatic repulsion.

Fundamental and non-fundamental particles

AQA A ▷ 1
AQA B ▷ 1
WJEC ▷ 2

> In an inelastic collision, kinetic energy is not conserved.

Electrons are **fundamental** particles, they cannot be split up. Protons and neutrons are themselves made up of other particles so they are not fundamental. Evidence for the structure of protons and neutrons comes from **deep inelastic scattering** experiments. In these experiments, very-high-energy electrons are fired at nucleons. The electrons are not affected by the strong force and they penetrate the nucleons in an inelastic collision, resulting in the electrons being scattered through a range of angles, with some being back-scattered in the same way that alpha particles can be back-scattered by gold foil.

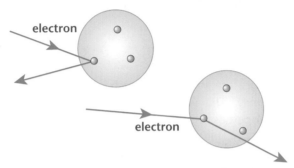

Deep inelastic scattering experiments show that nucleons contain small regions of intense charge.

This gives evidence that a nucleon contains small, dense regions of charge that are themselves fundamental particles. These particles are called **quarks**. The diagram above illustrates some of the findings of deep inelastic scattering. In the diagram below each nucleon is shown as containing three quarks.

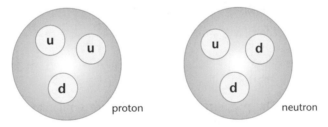

The quarks contained in nucleons

> The symbol **e** is used throughout this section to represent the amount of charge on an electron, ignoring the sign. Positive (+) and negative (−) show the sign of a charge.

Of the six types of quark, two occur in protons and neutrons. These are the **up** and **down** quarks. The up quark, symbol u, has a charge of $+\frac{2}{3}$e and that on the down quark, symbol d, is $-\frac{1}{3}$e. The diagram above shows the quarks in a proton and a neutron.

Particles and their antis

AQA A ▷ 1
AQA B ▷ 1
WJEC ▷ 2

> Where a particle has zero charge, for example the neutron, its antiparticle also has zero charge.

For every type of particle there is an antiparticle. An antiparticle:

- has the same mass as the particle, but has the opposite charge if it is the antiparticle of a charged particle
- annihilates its particle when they collide, the collision resulting in energy in the form of gamma radiation or the production of other particles. This process is called **pair annihilation**
- can be created, along with its particle, when a gamma ray passes close to a nucleus. This process is called **pair production**.

Anti-electrons, or **positrons**, are emitted in some types of radioactive decay and can be created when cosmic rays interact with the nuclei of atoms. Like all antiparticles, they cannot exist for very long on Earth before being annihilated by the corresponding particles.

The symbols β⁻ and β⁺ are normally used to refer to electrons and positrons given off as a result of radioactive decay.

> **Using symbols for particles**
> An electron is represented by the symbol e⁻ or β⁻.
> The positron can be represented by either of the symbols e⁺ and β⁺.
> The proton and antiproton are written as p⁺ and p⁻.
> More generally, a line drawn above the symbol for a particle refers to an antiparticle, so u and ū (read as u-bar) refer to the up quark and its antiquark.

KEY POINT

The greek letter ν is pronounced as 'new'.

The **antineutrino**, $\bar{\nu}$, is a particle of antimatter emitted along with an electron when a nucleus undergoes β⁻ decay. It is a particle that has almost no mass and no charge. Its corresponding particle the **neutrino**, ν, is emitted with a positron in β⁺ decay.

The quark family

AQA A   1
AQA B   1
WJEC   2

There are six quarks which, together with their antiquarks, make up the quark family. The name of each quark describes its type, or **flavour**. Some of the properties of quarks are:

- charge – this is $+\frac{2}{3}e$ or $-\frac{1}{3}e$ for quarks and is always conserved
- baryon number – like charge, this is conserved in all interactions
- strangeness – this property describes the strange behaviour of some particles that contain quarks. It is conserved in strong and electromagnetic interactions but can be changed in weak interactions.

The values of these properties are shown in the table.

The six quarks can be thought of as three pairs, u–d, c–s and t–b.

Quark	Symbol	Charge/e	Baryon number, B	Strangeness, S
up	u	$+\frac{2}{3}$	$\frac{1}{3}$	0
down	d	$-\frac{1}{3}$	$\frac{1}{3}$	0
charm	c	$+\frac{2}{3}$	$\frac{1}{3}$	0
strange	s	$-\frac{1}{3}$	$\frac{1}{3}$	−1
top	t	$+\frac{2}{3}$	$\frac{1}{3}$	0
bottom	b	$-\frac{1}{3}$	$\frac{1}{3}$	0

The properties of antiquarks are similar to those of the corresponding quark but with the opposite sign, so the antiquark $\bar{s}$ has charge $+\frac{1}{3}$, baryon number $-\frac{1}{3}$.

The hadrons

AQA A   1
AQA B   1
WJEC   2

The **hadrons** and **leptons** are two groups of particles. Hadrons are affected by both the strong and the weak nuclear force, but leptons are affected by the weak force only.

Hadrons are not fundamental particles; they are made up of quarks:
- **baryons** are made up of three quarks
- **mesons** consist of a quark and an antiquark.

The familiar baryons are the proton, whose quark structure is uud, and the neutron, udd. Protons are particularly stable – they rarely change into anything else. There are other baryons which have different combinations of quarks but they are all unstable and have short lifetimes. All baryons have a baryon number +1, as each quark contributes a baryon number of $+\frac{1}{3}$.

Two families of mesons are the **pi-mesons**, or **pions**, and the **kaons**. Because a meson consists of a quark and an antiquark, it has baryon number 0.

The table shows the structure and some properties of the kaons and pions.

Particle	Structure	Charge/e	Baryon number, B	Strangeness, S
π^0	$u\bar{u}$ or $d\bar{d}$	0	0	0
π^+	$u\bar{d}$	+1	0	0
π^-	$\bar{u}d$	−1	0	0
K^0	$d\bar{s}$	0	0	+1
K^+	$u\bar{s}$	+1	0	+1
K^-	$\bar{u}s$	−1	0	−1

The leptons

AQA A   1
AQA B   1
WJEC   2

Leptons are fundamental particles. There are three families of leptons:

- the electron and its antiparticle the positron, e^- and e^+, together with the neutrinos given off in β^- and β^+ decay. These neutrinos are called the electron-antineutrino, $\bar{\nu}_e$, and the electron-neutrino, ν_e, to distinguish them from the neutrinos associated with the other leptons
- the muons, μ^- and μ^+, together with their neutrinos ν_μ and $\bar{\nu}_\mu$. Muons have 200 times the mass of electrons, but are unstable, decaying to an electron and a neutrino
- the tauons, τ^- and τ^+ and their neutrinos ν_τ and $\bar{\nu}_\tau$. The tauons are the most massive leptons, having twice the mass of a proton. Like the muons, they are unstable.

Muons are produced by cosmic rays. Their properties are similar to those of electrons and positrons but they are much more massive. Tauons are the elephants of the lepton tribe; they are produced by the interaction of high-energy electrons and positrons.

Exchanging particles

AQA A   1
AQA B   1

The concept of a field is used to describe the effects of gravitational and electromagnetic forces and how they vary with distance. This is a useful concept that enables the size and direction of forces to be calculated, but it does not explain what causes them.

The interaction between particles that results in attractive and repulsive forces is due to a continual exchange of other particles. These exchange particles have a very short lifetime and owe their existence to borrowed energy, so they are often referred to as virtual particles.

There are different exchange particles associated with each of the four fundamental forces. For electromagnetic forces the exchange particles are photons, short bursts of electromagnetic radiation.

When two electrons approach each other they exchange virtual photons. The closer the electrons are to each other, the shorter the wavelength of the virtual photons exchanged.

A photon is a 'packet' of electromagnetic radiation. If you are unfamiliar with the concept of a photon, refer to section 4.3.

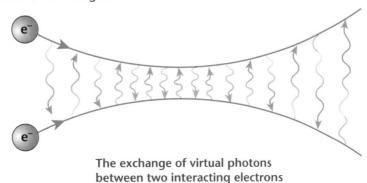

The exchange of virtual photons between two interacting electrons

Richard Feynman was an American physicist. He is famous for his books *Lectures on Physics* and for his way of explaining Physics concepts by using diagrams.

This exchange of virtual photons results in the electrons moving away from each other in the same way as two ice skaters on a collision course would move away from each other if they were to keep throwing things at each other!

The diagram on page 121 represents the exchange of virtual photons between two interacting electrons. Interactions between particles can be represented by **Feynman diagrams**. In these diagrams:

* straight lines are used for the interacting particles
* wavy lines represent the exchange particles
* the straight lines extend beyond the diagram, showing the existence of these particles before and after the interaction
* the wavy lines are contained within the diagram, showing that the exchange particles are short-lived.

This is a Feynman diagram for the interaction between two electrons shown in the diagram on page 121. The arrows do not represent directions of travel: arrows pointing in show the particles before the interaction and those pointing out show the particles after the interaction. The symbol γ is used to represent the exchange of virtual photons.

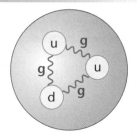

The exchange particles for gravitational forces are called **gravitons**, symbol g, though these have not yet been observed.

The strong force

AQA A   1
AQA B   1

The same symbol, g, is used for both gluons and gravitons.

It is the strong force that holds nucleons together in a nucleus and holds quarks together in a nucleon.

The exchange particles between quarks are called **gluons**, symbol g. The quarks that make up a proton or neutron are constantly exchanging gluons as they interact with each other.

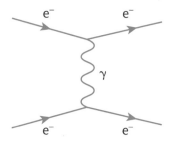

The exchange of gluons between the quarks in a proton

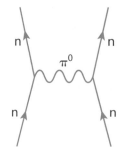 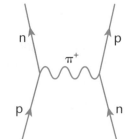 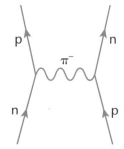

Interactions between nucleons

Pions, or pi-mesons, are responsible for the strong force between nucleons. Two protons or two neutrons exchange pi-zero, π^0, particles, but any one of the three pions can be exchanged during an interaction between a proton and a neutron. Feynman diagrams for some of the possible interactions are shown above.

Notice that charge and baryon number are conserved at each junction of the diagrams.

The weak force

AQA A   1

AQA B   1

The intermediate vector bosons form another group of particles, separate from the hadrons and leptons.

Can you interpret these Feynman diagrams? Try to describe the interaction that each one represents.

Weak interactions involve the exchange of one of three particles called **intermediate vector bosons**. Like the pi-mesons, their symbols, W^+, W^- and Z^0 indicate the charge on each boson. The W^+ boson carries charge $+e$ and the W^- boson carries charge $-e$.

- The W^+ boson transfers charge $+e$; it is exchanged in an interaction between a neutrino and a neutron, resulting in an electron and a proton.
- The W^- boson is responsible for β^- decay of a neutron.
- Weak interactions where there is no transfer of charge involve the Z^0 boson.

Feynman diagrams for these interactions are shown below.

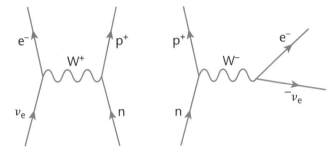

In two of these examples the type or flavour of a quark in the neutron has been changed from d to u by the weak interaction involving a charge-carrying boson. In β^- decay a down quark emits a W^- particle which decays to an electron and its antineutrino. This is shown in the diagram below.

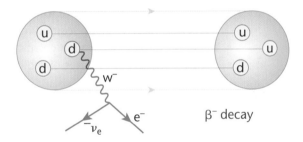

β^- decay

Progress check

1 State one piece of evidence for:
   a the particle nature of matter
   b the existence of nuclei.

2 Describe the differences between:
   a Hadrons and leptons
   b Baryons and mesons.

3 What change takes place in a proton in β^+ decay?

4 **a** What feature of the results of alpha-particle scattering experiments shows that the nucleus is positively charged?
   b Explain how the results of these experiments suggest that the nucleus occupies a very small proportion of the volume of the atom.

1 Possible answers include: **a** Brownian motion **b** alpha scattering

2 **a** Hadrons are made up of quarks.
   Leptons are fundamental particles.
   b Baryons contain 3 quarks. Mesons contain a quark and an antiquark.

3 A u quark changes to a d quark, changing the quark structure of the nucleon from uud to udd.

4 **a** The fact that some alpha particles are back-scattered shows that they are repelled from the nucleus, so they must have the same charge.
   b Most alpha particles pass through thin gold foil without being deviated.

4.2 Nuclei and radioactivity

After studying this section you should be able to:

- *explain that radioactivity is a natural phenomenon, involving emission of different types of ionising radiation*
- *describe the properties of alpha, beta-minus, beta-plus and gamma radiations*
- *apply rules of conservation of charge and nucleon number in writing balanced 'equations' to represent nuclear decay*
- *describe patterns of nuclear size and instability*

LEARNING SUMMARY

Radiation all around us

AQA A   1
AQA B   1

The term 'background radiation' is also used to describe the microwave radiation left over from the 'Big Bang'. This is a different type of radiation to nuclear radiation.

Radioactive decay occurs when an atomic nucleus changes to a more stable form. This is a random event that cannot be predicted. The emissions from these nuclei are collectively called radioactivity or radiation.

We are subjected to a constant stream of radiation called background radiation. Most of this is natural in the sense that it is not caused by the activities of people living on Earth. Sources of background radiation include:

- the air that we breathe – radioactive radon gas from rocks can concentrate in buildings
- the ground and buildings – all rocks contain radioactive isotopes
- the food that we eat – the food chain starts with photosynthesis. Radioactivity enters the food chain in the form of carbon-14, an unstable isotope of carbon that is continually being formed in the atmosphere
- radiation from space, called cosmic radiation
- medical and industrial uses of radioactive materials.

The emissions

AQA A   1
AQA B   1

The range of a beta-plus particle is effectively zero, since it is annihilated when it collides with an electron.

Alpha radiation is the most intensely ionising and can cause a lot of damage to body tissue. Although they cannot penetrate the skin, alpha emitters can enter the lungs during breathing.

Radioactive emissions are detected by their ability to cause ionisation, creating charged particles from neutral atoms and molecules by removing outer electrons. This results in a transfer of energy from the emitted particle, which is effectively absorbed when all its energy has been lost in this way. The four main emissions are alpha (α), beta-minus (β⁻), beta-plus (β⁺) and gamma (γ). Of these, alpha radiation is the most intensely ionising and has the shortest range, with the exception of beta-plus, while gamma radiation is the least intensely ionising and has the longest range.

- In alpha emission the nucleus emits a particle consisting of two protons and two neutrons. This has the same make-up as a helium nucleus.
- Beta-minus emission occurs when a neutron decays into a proton, emitting an electron in the process.
- In beta-plus emission a proton changes to a neutron by emitting a positron.
- A gamma emission is short-wavelength electromagnetic radiation.

Some properties of these radiations are shown in the table overleaf.

Radioactive emission	Nature	Charge/e	Symbol	Penetration	Causes ionisation	Affected by electric and magnetic fields
alpha	two neutrons and two protons	+2	^4_2He or $^4_2\alpha$	absorbed by paper or a few cm of air	intensely	yes
beta-minus	high-energy electron	–1	$^0_{-1}e$ or $^0_{-1}\beta$	absorbed by 3 mm of aluminium	weakly	yes
beta-plus	positron (antielectron)	+1	$^0_{+1}e$ or $^0_{+1}\beta$	annihilated by an electron	yes	yes
gamma	short-wavelength electromagnetic radiation	none	γ	reduced by several cm of lead	very weakly	no

Notice that the electron emitted in beta-minus decay and the positron emitted in beta-plus decay have been allocated the atomic numbers –1 and +1. This is because of the effect on the nucleus when these particles are emitted.

Balanced equations

Nucleon is a name for protons and neutrons (just as 'parent' is a name for mothers and fathers). An alpha particle has four nucleons – two protons and two neutrons – so its nucleon number, A, is 4. An electron has no nucleons, and its nucleon number is 0.

When a nucleus decays by alpha or beta emission, the numbers of protons and neutrons are changed. Gamma emission does not change the make-up of the nucleus, but corresponds to the nucleus losing excess energy. Gamma emission often occurs alongside alpha and beta emissions, though some artificial radioactive isotopes emit gamma radiation only.

The changes that take place due to alpha and beta emissions are:

* **alpha** – the number of protons decreases by two and the number of neutrons also decreases by two
* **beta-minus** – the number of neutrons decreases by one and the number of protons increases by one
* **beta-plus** – the number of neutrons increases by one and the number of protons decreases by one.

In writing equations that describe nuclear decay, both charge (represented by Z) and the number of nucleons (represented by A) are conserved. The table below summarises these changes and gives examples of each type of decay.

Check that the equations given as examples are balanced in terms of charge and number of nucleons.

Particle emitted	Effect on A	Effect on Z	Example
alpha	–4	–2	$^{226}_{88}Ra \rightarrow {}^{222}_{86}Rn + {}^4_2He$
beta-minus	unchanged	+1	$^{14}_{6}C \rightarrow {}^{14}_{7}N + {}^0_{-1}e$
beta-plus	unchanged	–1	$^{11}_{6}C \rightarrow {}^{11}_{5}B + {}^0_{+1}e$

Stable and unstable nuclei

Remember, isotopes of an element all have the same number of protons but different numbers of neutrons in the nucleus.

For stable nuclei with more than 20 protons, the neutron:proton ratio increases steadily to a value of around 1.5 for the most massive nuclei.

Carbon-11, carbon-12 and carbon-14 are three isotopes of carbon. Of these, only carbon-12 is stable. It has equal numbers of protons and neutrons. The graph shows the relationship between the number of neutrons (N) and the number of protons (Z) for stable nuclei.

It can be seen from this graph that the condition for a nucleus to be stable depends on the number of protons (Z):

* for values of Z up to 20, a stable nucleus has equal numbers of protons and neutrons
* for values of Z greater than 20, a stable nucleus has more neutrons than protons.

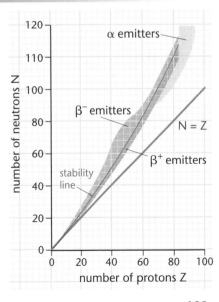

Unstable nuclei above the stability line in the graph are **neutron-rich**; they can become more stable by decreasing the number of neutrons. They decay by β^- emission; this leads to one less neutron and one extra proton and brings the neutron–proton ratio closer to, or equal to, one. An example is:

$$^{24}_{11}\text{Na} \rightarrow \,^{24}_{12}\text{Mg} + \,^{0}_{-1}\text{e}$$

Unstable nuclei below the stability line decay by β^+ emission; this increases the neutron number by one at the expense of the proton number. An example is:

$$^{11}_{6}\text{C} \rightarrow \,^{11}_{5}\text{B} + \,^{0}_{+1}\text{e}$$

The emission of an alpha particle has little effect on the neutron–proton ratio for isotopes that are close to the N = Z line and is confined to the more massive nuclei. For these nuclei, emission of an alpha particle changes the balance of the proton–neutron ratio in the favour of the neutrons. In the decay of thorium-228 shown below, the neutron–proton ratio increases from 1.53 to 1.55.

$$^{228}_{90}\text{Th} \rightarrow \,^{224}_{88}\text{Ra} + \,^{4}_{2}\text{He}$$

A carbon-12 nucleus consists of six protons and six neutrons. The mass of the atom is precisely 12 u, by definition, so after taking into account the mass of the electrons, that of the nucleus is 11.9967 u. The mass of the constituent neutrons and protons is:

$$6\,m_p + 6\,m_n = 6(1.0073\text{ u} + 1.0087\text{ u}) = 12.0960\text{ u}$$

The nucleus has less mass than the particles that make it up. This appears to contravene the principle of conservation of mass. Einstein established that *energy has mass*. The mass that you gain due to your increased energy when you walk upstairs is infinitesimally small, but at a nuclear level this mass cannot be ignored. Any change in energy is accompanied by a change in mass, and vice versa.

When dealing with the nucleus and nuclear particles, energy and mass are so closely linked that their equivalence, and that of their units, has been established:

> **KEY POINT**
>
> 1 u = 930 MeV
>
> where 1 eV (one electronvolt) is the energy transfer when an electron moves through a potential difference of 1 volt.

Using this relationship, the separate conservation rules regarding mass and energy can be combined into one so that (mass + energy) is always conserved in nuclear interactions.

To split a nucleus up into its constituent nucleons requires energy. It follows that a nucleus has less energy than the sum of the energies of the corresponding number of free neutrons and protons. So the fact that a nucleus has less energy than its nucleons would have in isolation, means that it also has less mass.

> **KEY POINT**
>
> The difference between the sum of the masses of the individual nucleons and the mass of the nucleus is called the **mass defect** or **nuclear binding energy**. It represents the energy required to separate a nucleus into its individual nucleons.

In the case of carbon-12 the mass defect, or nuclear binding energy, is equal to 0.093 u = 89.3 MeV.

As would be expected, the greater the number of nucleons, the greater the binding energy. The figure shows how the **binding energy per nucleon** varies with nucleon number. The most stable nuclei have the greatest binding energy per nucleon.

Side notes:

The N = Z line corresponds to a neutron–proton ratio of 1. As an alpha particle consists of two neutrons and two protons, its emission would hardly affect a neutron–proton ratio that is nearly 1.

Einstein's equation $E = mc^2$ gives the method for working out how much mass is associated with energy. Try using it to work out the increase in your mass when you walk upstairs.

An electronvolt is a tiny amount of energy compared to the joule, which is too large a unit to use on an atomic scale. $1\text{ eV} = 1.60 \times 10^{-19}\text{ J}$.

The greater the binding energy per nucleon, the more energy (per nucleon) is required to split the nucleus up, giving it more stability.

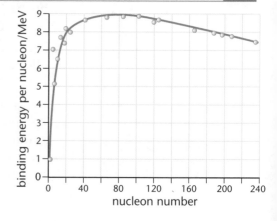

Energy spectra

AQA B ▸ 1

When a nucleus decays, the daughter nucleus has less mass and so more binding energy than the parent. The energy difference is the **decay energy**, the energy released by the nucleus as a result of its decay. This energy is transferred to:

> A nucleus can only decay naturally to a state where it has less energy.

- kinetic energy of the decay products
- energy of a gamma ray, if one is emitted
- energy of a neutrino or antineutrino, in the case of β emission.

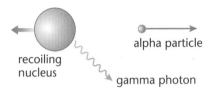

Energy being removed from a nucleus in alpha emission

recoiling nucleus

alpha particle

gamma photon

> The kinetic energy of the decay products includes that of the recoiling nucleus. This is most significant in the case of alpha decay.

There are different patterns in the energies of alpha, beta and gamma emissions.

- The energy of alpha particles depends on the source; some sources emit alpha particles that all have the same energy, other sources emit alpha particles with two or more possible energies.
- The energy of beta particles varies continuously over a range from zero up to a maximum value that depends on the source.
- In gamma emission, each source emits a line spectrum; the photons have only one or a small number of energies.

> The energies of the emitted particles are different for different sources. The shorter the half-life, the greater the energy of the emissions.

These cloud chamber tracks are from a source that emits alpha particles of two separate energies.

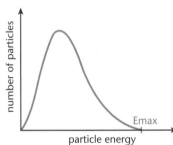

number of particles

particle energy

Emax

The range of energies in beta emission

While alpha and gamma emissions correspond to certain energy values, in beta emission most of the particles are emitted with less than the maximum energy available. This observation led to the discovery of the neutrino, v, and antineutrino, $\bar{v}$. The antineutrino is a particle emitted along with the electron in β^- decay and a neutrino is emitted along with a positron in β^+ decay. The energies of these particles account for the difference between the energy released by the nucleus and the kinetic energies of the beta particle and the recoiling nucleus together with the energy of any gamma ray photon that is emitted.

> Remember that gamma emission is usually due to a nucleus being left in an excited state following the emission of an alpha or beta particle.

Progress check

1 Tritium, ^3_1H, is a radioactive isotope of hydrogen. Use the N–Z graph to explain how it decays and to write an equation for the decay.

2 Strontium-90, $^{90}_{38}Sr$, decays to yttrium, symbol Y, by beta-minus emission. Write a balanced nuclear equation for this decay.

3 a In what way is the energy spectrum of beta emissions different to that of alpha and gamma emissions?

   b Explain how this is accounted for.

   b This is explained by the energy of the neutrino or antineutrino that accompanies a beta particle.

3 a The beta emission spectrum is continuous, whereas those of alpha and gamma only contain certain values of energy.

2   $^{90}_{38}Sr \rightarrow ^{90}_{39}Y + ^{0}_{-1}e$

$^3_1H \rightarrow ^3_2He + ^{0}_{-1}e$

1   As tritium lies above the N–Z line, it decays by beta-minus emission.

4.3 Light and matter

LEARNING SUMMARY

After studying this section you should be able to:

- describe evidence for quantised transfer of energy by light
- explain why gases have spectra of discrete wavelengths
- describe stimulated emission in a laser
- explain what is meant by wave–particle duality

The photoelectric effect

AQA A	1	Edexcel context	2
AQA B	1	OCR A	2
CCEA	2	OCR B	2
Edexcel concept	2	WJEC	2

Light and matter interact. Matter emits and absorbs electromagnetic radiation, losing energy to radiation or taking energy away from it. These processes tell us a lot about the nature of the world. One type of absorption (or transfer of energy from light to matter) that helped to change our view of the nature of light is the photoelectric effect.

The **photoelectric effect** provides evidence that electromagnetic waves have a particle-like behaviour which is more pronounced at the short-wavelength end of the spectrum. In the photoelectric effect electrons are emitted from a metal surface when it absorbs electromagnetic radiation.

The results of photoelectricity experiments show that:

- there is no emission of electrons below a certain frequency, called the **threshold frequency**, f_0, which is different for different metals
- above this frequency, electrons are emitted with a range of kinetic energies up to a maximum, $(\frac{1}{2}mv^2)_{max}$
- increasing the frequency of the radiation causes an increase in the maximum kinetic energy of the emitted electrons, but has no effect on the photoelectric current, i.e. the rate of emission of electrons
- increasing the intensity of the radiation has no effect if the frequency is below the threshold frequency; for frequencies above the threshold it causes an increase in the photoelectric current, so more electrons are emitted per unit time.

> The **threshold wavelength**, λ_0, is the wavelength of the waves that have the threshold frequency:
> $$\lambda_0 = c \div f_0$$

> The photoelectric effect can be demonstrated using a zinc plate connected to a gold leaf electroscope. An ultraviolet lamp discharges a negatively charged plate but has no effect on a positively charged plate.

The wave model cannot explain this behaviour; if electromagnetic radiation is a continuous stream of energy then radiation of all frequencies should cause photoelectric emission, it should only be a matter of time for an electron to absorb enough energy to be able to escape from the attractive forces of the positive ions in the metal.

> The word quantum refers to the smallest amount of a quantity that can exist. A quantum of electromagnetic radiation is the smallest amount of energy of that frequency.

The explanation for the photoelectric effect relies on the concept of a **photon**, a quantum or packet of energy. We picture electromagnetic radiation as short bursts of energy, the energy of a photon depending on its frequency.

A lamp emits random bursts of energy. Each burst is a photon, a quantum of radiation.

KEY POINT

The relationship between the energy, E of a photon, or quantum of electromagnetic radiation, and its frequency, f, is:
$$E = hf$$
where h is Planck's constant and has the value 6.63×10^{-34} J s.

The energy of a photon can be measured in either joules or electronvolts. The electronvolt is a much smaller unit than the joule.

The conversion factor for changing energies in eV to energies in joules is:
1.60 × 10⁻¹⁹ J eV⁻¹

> **KEY POINT**
>
> One electronvolt (1 eV) is the energy transfer when an electron moves through a potential difference of 1 volt.
> $$1 \text{ eV} = 1.60 \times 10^{-19} \text{ J}$$

Einstein's explanation of photoelectric emission is:

The work function is the **minimum** energy needed to liberate an electron from a metal. Some electrons need more than this amount of energy.

Radiation below the threshold frequency, f_0, no matter how intense, does not cause any emission of electrons.

- an electron needs to absorb a minimum amount of energy to escape from a metal. This minimum amount of energy is a property of the metal and is called the work function, ϕ
- if the photons of the incident radiation have energy hf less than ϕ, then there is no emission of electrons
- emission becomes just possible when hf = ϕ
- for photons with energy greater than ϕ, the electrons emitted have a range of energies, those with the maximum energy being the ones that needed the minimum energy to escape
- increasing the intensity of the radiation increases the number of photons incident each second. This causes a greater emission of electrons, but does not affect their maximum kinetic energy.

> **KEY POINT**
>
> Einstein's photoelectric equation relates the maximum kinetic energy of the emitted electrons to the work function and the energy of each photon:
> $$hf = \phi + \left(\tfrac{1}{2}mv^2\right)_{max}$$

At the threshold frequency, the minimum frequency that can cause emission from a given metal, $\left(\tfrac{1}{2}mv^2\right)_{max}$ is zero and so the equation becomes $hf_0 = \phi$.

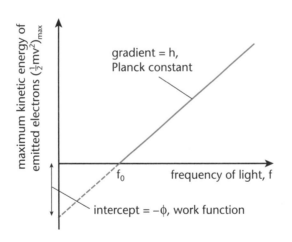

gradient = h, Planck constant

f_0

frequency of light, f

intercept = $-\phi$, work function

maximum kinetic energy of emitted electrons $\left(\tfrac{1}{2}mv^2\right)_{max}$

The equation:
$$hf = \phi + \left(\tfrac{1}{2}mv^2\right)_{max}$$
can be rewritten as:
$$\left(\tfrac{1}{2}mv^2\right)_{max} = hf - \phi$$
This is a graph of the form:
$$y = mx + c$$
where y is $\left(\tfrac{1}{2}mv^2\right)_{max}$
m is h
x is f
c is $-\phi$
A graph of the form $y = mx + c$ is always a straight line with gradient m and intercept c.

The word often used to describe separate distinct values, rather than continuous variation, is 'discrete'.

> **KEY POINT**
>
> The most important point about the photoelectric effect is that it tells us that light does not transfer energy smoothly and continuously, but in separate 'packets' or quanta. A photon is a name for a quantum of energy carried by light.

Line spectra

AQA A	1	Edexcel context	2
AQA B	1	OCR A	2
CCEA	2	OCR B	2
Edexcel concept	2	WJEC	2

With the exception of gamma rays, the emission of electromagnetic radiation by matter is associated with electrons losing energy. A hot solid can radiate a wide range of wavelengths through the infra-red and part of the visible spectrum, the extent of the range depending on its temperature.

When an electric current is passed through an ionised gas, only radiation with a small number of specific wavelengths is emitted. The wavelengths are characteristic of the gas used and are called a **line spectrum**.

Spectra can be seen by isolating a gas at a low pressure in an enclosed glass container, and applying a potential difference to it to supply energy. The light that the gas emits is allowed to escape from the container through a narrow slit, to act as a small and coherent source. A diffraction grating separates the different wavelengths of light that the gas emits, and because the light comes from a narrow slit, the spectrum appears as a set of lines parallel to the slit. Different gases have different spectra. The diagram below shows part of a hydrogen spectrum. The same pattern of wavelengths, or spectrum, is seen with all samples of hydrogen.

The diagram shows some of the lines present in the hydrogen spectrum and their wavelengths.

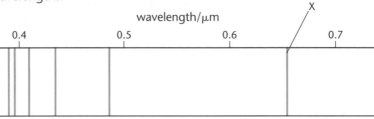

> Not all of these lines lie in the visible spectrum, which extends from 0.4 μm to 0.65 μm.

The existence of line spectra provides evidence that the electrons in orbit around a nucleus can only have certain values of energy, the values being characteristic of an atom. Energy can only be emitted or absorbed in amounts that correspond to the differences between these allowed values.

An **energy level diagram** shows the amounts of energy that an electron can have. The diagram below is an energy level diagram for hydrogen. Note that on an energy level diagram:

> In some circumstances, radiation from solids can include line spectra. In an **X-ray tube**, high energy electrons hit a solid target. The rapid deceleration of electrons produces radiation with a continuous spectrum, but superimposed on this are peaks with specific wavelengths that are characteristic of the material that the target is made of.

- the energies are measured relative to a zero that represents the energy of an electron at rest outside the atom, i.e. one that is just free
- an orbiting electron has less energy than a free electron, so it has negative energy relative to the zero

> Line spectra are observed for all elements, though of course many have to be heated to become gases.

- an electron with the minimum possible energy is in the **ground state**; higher energy levels are called **excited states**.

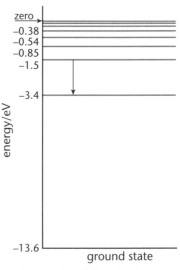

Energy levels in a hydrogen atom

> Movement of an electron to the ground state results in the emission of a photon with an energy in excess of 10 eV. Photons with this amount of energy give spectral lines in the ultraviolet region of the spectrum.

An electron movement (or transition) between the two levels as shown by the arrow in the diagram always produces light of a particular wavelength, of just over 0.65 μm. This corresponds to line X shown in the spectrum above.

> The lines of the emission spectrum often appear black in the absorption spectrum. The photons of these energies are absorbed as the electrons move to more excited states, and then released when the electrons lose energy. The emitted photons are radiated in all directions, so very little energy is detected in any one direction.

Energy is emitted in the form of a photon when an electron moves from an excited state to a lower energy level. The energy of the photon is equal to the difference in the values of the energy levels. For example, an electron moving from an energy level with −0.38 eV of energy to one with −0.85 eV loses energy equal to (−0.38 − −0.85) eV = 0.47 eV. This corresponds to emitting a photon of frequency 1.13×10^{14} Hz, which lies in the infra-red part of the electromagnetic spectrum.

> **KEY POINT**
>
> When an electron moves from an energy level E_1 to a lower energy level E_2, the energy (hf) of the photon emitted is given by:
> $$hf = E_1 - E_2$$

Electrons can gain energy by absorbing photons. As with emission, the only photons that can be absorbed are those that correspond to allowed movements, or transitions, of electrons. An **absorption spectrum** is produced by shining white light through a sample of a gaseous element. The spectrum that emerges is the full spectrum with the element's emission spectrum missing or of low intensity. This is due to the electrons absorbing photons of just the right energy to allow them to move to a more excited state.

Ionisation

AQA A	1	OCR A	2
AQA B	1	OCR B	2
CCEA	2	WJEC	2
Edexcel context	2		

Note that because a hydrogen atom is so simple, a hydrogen ion is just a proton.

Hydrogen atoms are the smallest and simplest, with one proton in the nucleus and with just one electron. A hydrogen atom in its ground state has its electron in the lowest possible energy level. The electron would need 13.6 eV of energy in order to escape from the atom to create a free electron and a hydrogen ion. That is, 13.6 eV is needed to ionise the atom, so it is called **ionisation energy**.

In a fluorescent lighting tube, a high potential difference causes some ionisation, and the ions and the freed electrons then accelerate in opposite directions. This causes violent collisions, and more ionisation as well as excitation of atoms. When electrons return towards the ground state they lose energy, which is emitted as electromagnetic radiation. Domestic lamps use mercury as the active gas inside the tube, and this emits UV radiation. The white coating inside the tube absorbs the UV and emits visible light.

Lasers

CCEA	2
WJEC	2

Laser devices are very common – the very coherent light source is excellent for reading information from surface patterns such as bar codes and CD pits. The word laser comes from Light Amplification by Stimulated Emission of Radiation.

A sample of material in a laser is deliberately excited, or given extra energy so that large numbers of electrons move out of the ground state to higher energy levels. That reverses the normal situation, where most electrons are in the ground state, to one in which most atoms are excited. It is called **population inversion**.

The return of electrons to the ground state results in the emission of light. 'Amplification' occurs because a photon released by one such emission process can stimulate emission of an identical photon in an encounter with another atom in its excited state. Mirrors can hold photons within the material to increase their chances of stimulating further emission. The result is the rapid rate of emission of photons, for as long as energy continues to be provided to maintain atoms in their excited states. The process of supplying the energy is called **pumping**.

Particles or waves

AQA A	1	Edexcel context	2
AQA B	1	OCR A	2
CCEA	2	OCR B	2
Edexcel concept	2	WJEC	2

Try calculating the de Broglie (pronounced de Broy) wavelength of a moving snooker ball. Is it possible for the ball to show wave-like behaviour?

If waves can show particle-like behaviour in photoelectric emission, can particles also behave as waves? Snooker balls bounce off cushions in the same way that light bounces off a mirror, so reflection is not a test for wave-like or particle-like behaviour. Diffraction and interference are properties unique to waves, so particles can be said to have a wave-like behaviour if they show these properties.

All particles have an associated wavelength called the **de Broglie** wavelength.

> The wavelength, λ, of a particle is related to its momentum, mv, by the de Broglie equation:
> $$\lambda = h/mv$$
> where h is the Planck constant.
>
> **KEY POINT**

The de Broglie wavelength of such an electron is of the order of 1×10^{-10} m.

An electron that has been accelerated through a potential difference of a few hundred volts has a wavelength similar to that of X-rays and gamma rays.

This wavelength is also similar to the spacing of the atoms in crystalline materials, so these materials provide suitable-sized gaps to cause diffraction.

You may have seen this pattern formed on a fluorescent screen in a vacuum tube.

Diffraction patterns formed by a beam of electrons after passing through thin foil or graphite show a set of bright and dark rings on photographic film, similar to those formed by X-ray diffraction.

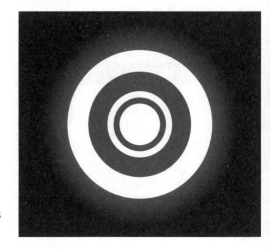

A diffraction pattern formed by passing a beam of electrons through graphite

Electrons can also be made to interfere when two coherent beams overlap. They produce an interference pattern similar to that of light, but on a much smaller scale.

Other particles such as protons and neutrons also show wave-like behaviour.

Particles and waves are the models that we use to describe and explain physical phenomena. It is not surprising that the real world does not fit neatly into our models.

There are two separate models of how matter behaves. The particle model explains such phenomena as ionisation and photoelectricity, while the wave model explains interference and diffraction. It is not appropriate to classify matter as waves or particles because photons and electrons can fit either model, depending on the circumstances.

> The use of two models to think about light and matter at small scales is called **wave–particle duality**.
>
> KEY POINT

Progress check

1 A photon of green light has a wavelength of 5.0×10^{-7} m.
   a Calculate the frequency of the light.
   b Calculate the energy of the photon:
      i in J
      ii in eV.

2 Calculate the de Broglie wavelength of an electron, $m_e = 9.1 \times 10^{-31}$ kg, travelling at a speed of 3.0×10^7 m s^{-1}.

3 An electron in a hydrogen atom undergoes a transition from an energy level of -0.54 eV to one of -3.40 eV.
   a Calculate the frequency of the photon emitted.
   b In what part of the electromagnetic spectrum is the emitted radiation?

3 a 6.90×10^{14} Hz b visible (light)
2 2.43×10^{-11} m
1 a 6.0×10^{14} Hz b i 3.98×10^{-19} J ii 2.49 eV

4.4 The wider Universe

After studying this section you should be able to:

- *explain how we know about stars' chemical compositions, temperature and luminosity*
- *describe how information on temperature and luminosity is used to classify stars*
- *explain why stars emit photons and neutrinos*
- *explain what red shift is and why it is important*
- *explain how we can estimate the age of the Universe*
- *consider the challenge of dark matter*

Stars' absorption spectra

Stars are a long way away, and all we have is their light by which to study them. The first thing we can do is to disperse the light into spectra. That shows a continuous spectrum with a peak of intensity, and with some discrete colours being absent.

Take the absent colours first. These appear as dark lines across the spectrum. They are absorption lines, and their patterns provide matches with absorption lines seen by shining light through different materials here on Earth. The absorption takes place as light passes through material in the outer layers of stars at the start of its long journey into space. The patterns, matched with those seen on Earth, show us what chemical elements those outer layers are made of. So light from a star tells us, in detail, about the chemical composition of its outer layers.

Stars' continuous spectra

A perfect black body is one for which the emission of light depends only on temperature, and not on other factors such as shininess or reflecting properties. Stars behave much like black bodies.

Experiments on black bodies here on Earth show that their spectra have particular patterns, with a peak of intensity that is related to temperature. The relationship is called **Wien's law**, and it can be written as:

temperature, T, in Kelvin (K) = $2.9 \times 10^{-3}/\lambda_{peak}$

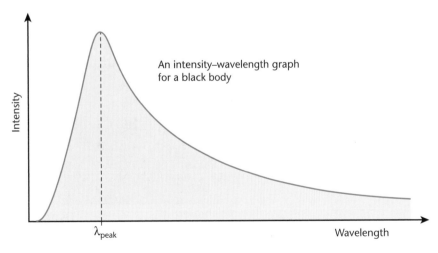

An intensity–wavelength graph for a black body

So, from the value of wavelength at which a star's spectrum peaks, we can find out the surface temperature of the star.

Intensity of emission

AQA B   1
WJEC   2

Note that comparison of the intensity of light from a star that we see here on Earth with its intensity as known from its temperature gives us an indication of how far away it is, by application of the inverse square law, $I \propto 1/R^2$.

Laboratory observations of the rate of emission of energy per unit area of a body, its emission intensity in Wm^{-2}, is also related to temperature. The relationship this time is given by **Stefan's law**:

intensity of emission, in Wm^{-2} = constant × T^4

This is also useful, because it tells us the intensity with which a star radiates, without having to go there.

Luminosity of a star is the rate of emission of energy, measured in watts. It is the star's intensity of emission multiplied by its total surface area:

luminosity, in W = intensity of emission × star surface area

Star classification

AQA B   1
WJEC   2

Colour is related to temperature – just as it is for a body heated on Earth. At relatively low temperatures, a body glows dull red and its colour shifts towards the blue end of the spectrum as it gets hotter.

One way to classify stars is simply by their temperature, as known by examining their spectra and applying Wien's law. This gives a sequence that runs from class O, the hottest 'blue' stars with surface temperatures of around 30 000 K, to 'red' class M stars, with surface temperature of about 3000 K.

The full sequence of temperature classifications is shown on the **Hertzsprung–Russell diagram** below. (Note that temperature decreases from left to right along the temperature axis of the diagram.)

The Hertzsprung–Russell diagram is a graph onto which stars can be plotted, according to their temperature and luminosity. Stars with similar positions are regarded as stars of the same kind, and classified accordingly, and given names according to their colour (as indicated by temperature as well as direct appearance) and according to their size (as indicated by their luminosity). Thus there are blue giants, red giants, white dwarfs and red dwarfs.

The Sun is plotted near the centre of the diagram and and is a yellow star.

As stars change, or evolve, their positions on the diagram change.

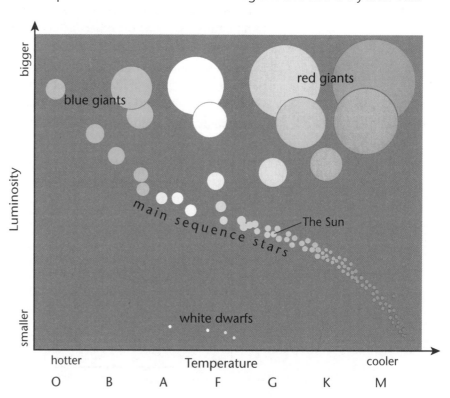

The Hertzsprung–Russell diagram – a visual classification of stars

The nuclear processes within stars

Stars gain their energy by converting mass to energy. The conversion factor is large (the square of the speed of light c, as given by $E = mc^2$) so small amounts of mass make large amounts of energy available. The processes involve interactions on the nuclear scale, including:

$$^1_1p + ^1_1p \rightarrow ^2_1d + ^0_1e^+ + \upsilon_e$$
$$^2_1d + ^1_1p \rightarrow ^3_2He + \gamma$$
$$^3_2He + ^3_2He \rightarrow ^4_2He + ^1_1p + ^1_1p$$

p = proton
d = deuterium

Overall:

$$4^1_1p \rightarrow ^4_2He + 2^0_1e^+ + 2\upsilon_e + 2\gamma$$

The process is a source of photons and neutrinos, which carry energy away from stars. Positrons are also produced, but these are antimatter and do not survive long since they are surrounded by matter.

Red shift and the expanding Universe

Further study of absorption spectra of distant light sources reveals a very significant observation. Other galaxies are moving away from our own, and the further away they are, the faster they are moving. The space around us is getting bigger.

This knowledge comes from the Doppler effect (see page 95). There is a shift in frequency of received light. The familiar patterns of absorption lines are shifted towards the red end of the spectrum, towards longer wavelength and lower frequency. It is this red shift that tells us that the Universe seems to be expanding. The size of the shift is given by:

$$\Delta f = f\, v/c$$

where f is the frequency of a dark absorption line in the laboratory, v is the relative speed of source and observer, which in this case, is called speed of recession, and c is the speed of light.

Discovery of the expansion of the Universe gave rise to the Big Bang theory. This is supported by other evidence, particularly by the existence of cosmic microwave background radiation. It is a theory that scientists are very much continuing to explore, looking for further support or for falsification.

The Big Bang idea supposes that the Universe – which means all of time and all of space – began from a single point, and went through phases of rapid expansion, or inflation, followed by the establishment of the four forces (gravitational, electric, strong and weak forces) and of particles like quarks, electrons, photons and neutrinos. A process of nucleosynthesis followed – the creation of protons and neutrons and of small nuclei, setting up a universal composition that we still see – a Universe of about 75% hydrogen and 24% helium.

For a long time, this soup was too hot for atoms to form – much of the energy of the Universe existed as photons, or radiation (of which the cosmic microwave background is the remains). Atoms did form eventually, though, and from initial random irregularities in their distribution the matter clumped together to make the first stars. Processes of change are still going on.

The Hubble law and the age of the Universe

> A megaparsec is a measure of distance, equal to about 3.26 million light years or about 3×10^{22} metres.

> Quasars are objects with very large red shift, suggesting that they are a long way away. Since light takes time to travel the light we receive from such distant objects must have left them billions of years ago. This, in turn, suggests that they are objects formed in the early Universe. Quasars are now believed to be bodies of matter surrounding supermassive black holes in the centres of young galaxies.

The statement that speed of recession is proportional to distance from us is called the Hubble law. The constant of proportionality is the Hubble constant.

speed of recession = Hubble constant × distance, or $v = Hd$

Speed of recession has been measured for many galaxies, in order to find out, as closely as possible, the value of the Hubble constant. The most reliable measurements available so far suggest that the value is $71 \pm 7\,\mathrm{km\,s^{-1}}$ megaparsec^{-1}, which equates to $2.3 \times 10^{-18}\,\mathrm{s^{-1}}$.

The value of Hubble's constant provides an estimate of the age of the Universe. Since the Big Bang theory suggests that everything in the Universe was once at a single point, the speed with which any two bodies have been moving apart since the beginning of the Universe is given by:

$v = d/T$, where T is the total time available, or the age of the Universe

We can combine this with the equation above:

$Hd = d/T$

which becomes:

$T = 1/H$

So if H is $71\,\mathrm{km\,s^{-1}}$ megaparsec^{-1}, then T is approximately 13.5 billion years. This is a crude estimate, based (for example) on an assumption that the Hubble constant has indeed always been constant, but it is the best we have.

Galaxies and dark matter

After all our work in Physics, it seems that there is still plenty of work to do. Observations of galaxies, for example, tell us that there is far more matter in the Universe than we have been able to detect directly until now – and we do not know what it is! Since it emits no light, it is called dark matter.

Evidence for its existence comes from observing the rotation of galaxies. They behave exactly as if they were much bigger than the combined masses of their visible stars. The scientists who solve this riddle will make their names.

Progress check

1 State the nature of the evidence that can show:
   a the chemical composition of a star
   b the surface temperature of a star.

2 Sketch intensity–wavelength graphs (spectra) for a red giant and a white dwarf, on the same axes.

3 A few galaxies whose light is blue-shifted do exist. Blue shifted absorption lines are shifted towards the blue end of the visible spectrum.
   a Explain how blue shift happens.
   b Explain why such galaxies must be relatively close to our own.

is dominated by the expansion; since rate of recession is greater at greater distance.
b The galaxies have some motion superimposed on the overall expansion; for further galaxies relative motion
3 a The relative motion of source and observer is towards each other.
2 red giant has higher peak at longer wavelength.
b The peak on the spectrum is at a wavelength that is related to temperature by Wien's law.
1 a The absorption spectra

Practice examination questions

1 The diagram shows some possible outcomes when alpha particles are fired at gold foil.

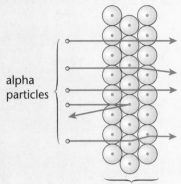

atoms in gold foil

(a) Explain how these outcomes show that:
  (i) the nucleus is positively charged [2]
  (ii) most of the atom is empty space. [1]

(b) In **deep inelastic scattering** electrons are used to penetrate the nucleus.
  (i) What is meant by 'deep inelastic scattering'? [2]
  (ii) Explain why electrons are able to probe further into the nucleus than alpha particles can. [2]
  (iii) Explain why the proton and the neutron are not fundamental particles. [1]
  (iv) How does deep inelastic scattering provide evidence for quarks? [1]
  (v) Describe the quark structure of a proton and a neutron. [2]

2 1.00 g of carbon is obtained from the wood of a living tree.
(a) Calculate the number of carbon atoms present in 1.00 g, assuming that all the carbon atoms are carbon-12.
The Avagadro constant, $N_A = 6.02 \times 10^{23}$ mol^{-1}.
The mass of 1 mole of carbon-12 is 12 g precisely. [2]

(b) Natural carbon consists of approximately:
99% carbon-12
1% carbon-13
1×10^{-10} % carbon-14.

  (i) Explain whether the actual number of carbon atoms in a 1.00 g sample is likely to be significantly different from the answer to (a). [2]
  (ii) Estimate the number of carbon-14 atoms present in the sample. [1]

3 For nuclei with up to 20 protons, a stable nucleus has approximately equal numbers of protons and neutrons.
(a) Explain how, for nuclei with less than 20 protons, the type of decay undergone by unstable nuclei depends on the neutron–proton ratio. [2]

(b) Describe how the decay of $^{13}_{6}C$ results in the formation of a more stable nucleus. [2]

(c) How does the neutron–proton ratio change for more massive stable nuclei? [2]

Practice examination questions (continued)

4 (a) How does the structure of a **meson** differ from that of a **baryon**? [2]

(b) The Feynman diagram illustrates the interaction between two protons.

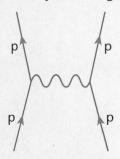

The exchange particle is a pion.
(i) Which pion is exchanged in this interaction? [1]
(ii) Explain why it is this pion rather than a different one. [2]

5 When electromagnetic radiation is incident on a clean surface of copper, *photoelectric emission* may take place. No emission occurs unless the photon energy is greater than or equal to the *work function*.

(a) Explain the meaning of the terms:
(i) photoelectric emission [2]
(ii) photon [2]
(iii) work function. [2]

(b) The work function of copper is 5.05 eV.
(i) Calculate the minimum frequency of radiation that causes photoelectric emission.
$e = 1.60 \times 10^{-19}$ C
$h = 6.63 \times 10^{-34}$ J s [2]
(ii) Calculate the maximum kinetic energy of the emitted electrons when electromagnetic radiation of frequency 1.80×10^{15} Hz is incident on the copper surface. [3]

Practice examination answers

Chapter 1 Force, motion and energy

1   The vector diagram is shown on the right [1].
    The resultant velocity is 19 m s^{-1} [1]
    at an angle of 18° E of N [1].

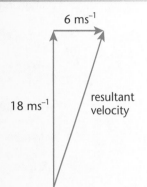

[Total: 3]

2   (a) 450 N × cos 50° [1] = 289 N [1]
    (b) 289 N [1]
    (c) The light fitting is in equilibrium [1] so the sum of the forces on it is zero [1].
    (d) The vector diagram is shown on the right [1].
         S = 345 N [1].

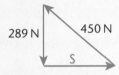

[Total: 7]

3   (a) Acceleration = change in velocity ÷ time [1] = 24 m s^{-1} ÷ 80 s [1] = 0.30 m s^{-2} [1].
    (b) Between 230 and 290 s [1].
    (c) (i)   20 s [1]
         (ii) 120 s [1].
    (d) 960 m + 1680 m + 720 m + 900 m + 1500 m + 900 m = 6660 m [2] (for all correct).
    (e) 3360 m − 3300 m = 60 m [1].   [Total: 9]

4   (a) Use s = ut + ½at^2 [1] t = $\sqrt{(2 \times 0.39 \text{ m} \div 10 \text{ m s}^{-2})}$ [1] = 0.28 s [1].
    (b) s = ut [1] = 35 m s^{-1} × 0.28 s = 9.8 m [1].
    (c) u = s ÷ t [1] = 15 m ÷ 0.28 s = 54 m s^{-1} [1].   [Total: 7]

5   (a) t = Δv ÷ a [1] = 14 m s^{-1} ÷ 2 m s^{-2} = 7 s [1].
    (b) distance = average speed × time [1] = 21 m s^{-1} × 7 s = 147 m [1].   [Total: 4]

6   (a) Use v^2 = u^2 + 2 as [1] u^2 = − 2 × − 10 m s^{-2} × 3.5 m = 70 m^{2}s^{-2} [1] u = 8.4 m s^{-1} [1].
    (b) Time travelling upwards, t = distance ÷ average speed = 3.5 m ÷ 4.2 m s^{-1} = 0.83 s [1]
         Time in air = 1.66 s [1].   [Total: 5]

7   (a) Use v^2 = u^2 + 2as [1] a = v^2 ÷ 2s = (60 m s^{-1})^2 ÷ 2 × 1500 m [1] = 1.2 m s^{-2} [1].
    (b) F = ma [1] = 70 000 kg × 1.2 m s^{-2} [1] = 84 kN [1].
    (c) As the aircraft gains speed the size of the resistive forces increases [1].
         This causes the resultant force to decrease [1].   [Total: 8]

8   (a) (i)   The pull of the trailer on the car [1]
         (ii) 150 N backwards [1].
    (b) (i)   130 N forwards [1]
         (ii) a = F ÷ m [1] = 130 N ÷ 190 kg [1] = 0.68 m s^{-2} [1].
    (c) (i)   0 [1]
         (ii) 150 N backwards [1].   [Total: 8]

9   (a) (i)   40 kg × g = 400 N [1]
         (ii) The child pulls the Earth [1].
    (b) (i)   400 N × cos 66° [1] = 163 N [1]
         (ii) a = F/m [1] = 73 N ÷ 40 kg [1] = 1.8 m s^{-2} [1].
    (c) Use v^2 = u^2 + 2as [1] v^2 = 2 × 1.8 m s^{-2} × 5.5 m = 19.8 m^2 s^{-2} [1] v = 4.4 m s^{-1} [1].
                                                                  [Total: 10]

Practice examination answers

10 (a) The astronaut gains momentum in the opposite direction to the push on the spacecraft [1]. The spacecraft gains the same amount of momentum, in the direction of the astronaut's push [1].
(b) The change in momentum of the gas [1] is balanced by an equal and opposite change in momentum of the astronaut [1].   [Total: 4]

11 (a) $W = F \times s$ [1] = 30 000 N × 18 m [1] = 540 kJ [1].
(b) 540 kJ [1]
(c) $P = W \div t$ = 540 kJ ÷ 24 s [1] = 22.5 kW [1].
(d) $P_{in} = P_{out} \div$ efficiency [1] = 22.5 kW ÷ 0.45 = 50 kW [1].   [Total: 8]

12 (a) $\Delta E_p = mg\Delta h$ [1] = 2.0 × 10⁵ kg × 10 N kg⁻¹ × 215 m [1] = 4.30 × 10⁸ J [1].
(b) $E_k = \frac{1}{2}mv^2$ = 4.30 × 10⁸ J [1] v = √(2 × 4.30 × 10⁸ J ÷ 2.0 × 10⁵ kg) [1] = 65.6 m s⁻¹ [1].
(c) Kinetic energy of water leaving turbines each second = $\frac{1}{2}mv^2 = \frac{1}{2}$× 2.0 × 10⁵ kg × (8 m s⁻¹)² = 6.4 × 10⁶ J [1]. Maximum power input = 4.30 × 10⁸ W – 6.4 × 10⁶ W = 4.24 × 10⁸ W [1].
(d) Efficiency = useful power out ÷ power in = 2.5 × 10⁸ W ÷ 4.24 × 10⁸ W [1] = 0.59 [1].   [Total: 10]

13 (a) Energy = $\frac{1}{2}Fx = \frac{1}{2}$ × 25 N × 0.15 m [1] = 1.875 J [1].
(b) (i) $\frac{1}{2}mv^2$ = 1.875 J [1] v² = 2 × 1.875 J ÷ 0.020 kg = 187.5 m² s⁻² [1] v = 13.7 m s⁻¹ [1]
(ii) $mg\Delta h$ = 1.875 J [1] Δh = 1.875 J ÷ 0.20 N [1] = 9.4 m [1].   [Total: 8]

14 (a) Moment = force × distance to pivot = 60 N × 0.45 m [1]
= 27 N m [1].
(b) F = 27 N m ÷ 0.06 m [1] = 450 N [1].
(c) A bigger moment is caused for the same applied force [1].
So the force acting on the branch is greater [1].   [Total: 6]

15 (a) Horizontal component = 4500 × cos 50 = 2893 N [1].
Vertical component = 4500 × cos 40 = 3447 N [1].
(b) pressure = normal force ÷ area [1] = 3447 N ÷ 0.09 m² [1] = 38300 Pa [1].
(c) To reduce the pressure on the ground [1] so that the girder does not penetrate the surface [1].   [Total: 7]

16 (a) Volume of water = 10.5 m × 4.2 m × 3.5 m = 154 m³ [1]
Mass = volume × density = 1.54 × 10⁵ kg [1]
Weight = mass × g = 1.54 × 10⁶ N [1].
(b) Pressure = normal force ÷ area [1] = 1.54 × 10⁶ N ÷ 4.41 × 10¹ m² [1] = 3.5 × 10⁴ Pa [1].
(c) Liquid pressure acts equally in all directions [1] so at the bottom of the tank the pressure on the sides is equal to that on the base [1].
(d) It is the same [1] because the pressure depends on the depth only [1].   [Total: 10]

17 (a) A ductile material undergoes plastic deformation [1] but a brittle material breaks [1].
(b) (i) 340 MPa [1]
(ii) up to 300 MPa [1]
(iii) E = stress ÷ strain = 3.00 × 10⁸ Pa ÷ 1.50 × 10⁻³ [1] = 2.00 × 10¹¹ Pa [1].
(iv) The curve would have the same shape [1]; the values on the axes would be the same as the stress–strain graph is the same for all samples of the material [1].   [Total: 8]

Chapter 2 Electricity

1 (a) $R = \rho l/A$ [1].

(b) Resistivity is a property of a material [1]; the resistance of an object also depends on its dimensions [1].

(c) Cross-sectional area, $A = 2.0 \times 10^{-5}$ m^2 [1] $R = 3.00 \times 10^{-5}$ Ω m $\times 1.0 \times 10^{-2}$ m ÷ 2.0×10^{-5} m^2 [1] $= 1.5 \times 10^{-2}$ Ω [1]. [Total: 5]

2 (a) $1/R = 1/10$ Ω $+ 1/10$ Ω [1] $R = 5$ Ω [1].

(b) $I = V/R$ [1] $= 6.0$ V ÷ 20 Ω [1] $= 0.30$ A [1].

(c) 0.30 A ÷ 2 [1] $= 0.15$ A [1].

(d) $V = IR = 0.15$ A × 10 Ω [1] $= 1.5$ V [1]. [Total: 9]

3 (a) (i) The graph should show the resistance decreasing with increasing temperature [1] in a non-linear way [1]

(ii) The resistance of a metallic conductor increases with increasing temperature [1] linearly [1]

(iii) Ohm's law only applies to metallic conductors [1] at constant temperature [1].

(b) (i) Current in circuit $= 9.0$ V ÷ 600 Ω $= 1.5 \times 10^{-2}$ A [1].
Thermistor p.d. $= 1.5 \times 10^{-2}$ A × 500 Ω [1] $= 7.5$ V [1]

(ii) The p.d. across the thermistor decreases [1] as it has a smaller proportion of the circuit resistance [1]

(iii) Current in circuit $= 9.0$ V ÷ 175 Ω $= 5.1 \times 10^{-2}$ A [1]. Resistor p.d. $= 5.1 \times 10^{-2}$ A × 100 Ω [1] $= 5.1$ V [1]

(iv) Current in circuit $= 3.9$ V ÷ 300 Ω $= 0.013$ A [1]. Resistor p.d. $= 5.1$ V [1]. Resistor value $= 5.1$ V ÷ 0.013 A $= 392$ Ω [1]. [Total: 17]

4 (a) The current in the cell and the digital voltmeter is negligible [1] so there is no voltage drop due to internal resistance [1]. With the moving coil voltmeter there is a significant current in the cell; the voltage drop is the p.d. across the internal resistance [1].

(b) $R = V/I$ [1] $= 0.05$ V ÷ 1.55×10^{-3} A [1] $= 32.3$ Ω [1]. [Total: 6]

5 (a) Current in each lamp, $I = P/V = 6$ W ÷ 12 V $= 0.50$ A. Current in battery $= 4 \times 0.50$ A $= 2.0$ A [1].

(b) $\Delta q = I\Delta t$ [1] $= 2.0$ A × 60 s [1] $= 120$ C [1].

(c) $R = V/I$ [1] $= 12$ V ÷ 2.0 A [1] $= 6.0$ Ω [1]. [Total: 7]

6 (a) (i) $I^2 = P/R$ [1] $I = \sqrt{(10 \text{ W} \div 4.7 \text{ Ω})}$ [1] $= 1.46$ A [1]

(ii) $V = IR = 1.46$ A × 4.7 Ω [1] $= 6.9$ V [1]

(iii) $I = AR/\rho$ [1] $= 2.0 \times 10^{-7}$ m^2 × 4.7 Ω ÷ 5.0×10^{-7} Ω m [1] $= 1.88$ m [1].

(b) A smaller length of material is needed [1] for the same resistance [1]. [Total: 10]

7 (a) Potential divider [1].

(b) Apply a varying p.d. to component X [1].

(c) Circuit B varies the current and only allows investigation of the characteristics when the diode is conducting [1]. Circuit A allows the diode to be investigated when it is not conducting and when it is conducting [1].

(d) (i) The completed table is:

resistance /Ω	10.8	3.55	1.73	1.15	0.80

[Two marks for all correct.]

(ii) The graph is shown on the right.
Marks are awarded for:
Scales and labelling of axes [1]
Correct plotting [2]
Drawing a smooth curve [1]

(iii) 0.785 V [1]

(iv) The diode only conducts over a narrow range of voltages [1].

[Total: 12]

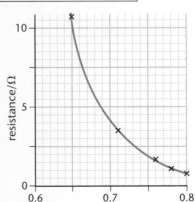

141

8 (a) $v = I/nAe$ [1] $= 5$ A $\div$ $(8.0 \times 10^{28}$ m$^{-3} \times 1.25 \times 10^{-6}$ m$^2 \times 1.60 \times 10^{-19}$ C) [1] $= 3.1 \times 10^{-4}$ m s^{-1} [1].

(b) The cross-sectional area of the tungsten is smaller than that of the copper [1] and the concentration of free electrons is less [1].

(c) Heating is due to collisions between the free electrons and the metal ions [1]. The collisions are more frequent in the tungsten than in the copper [1]. [Total: 7]

9 (a) Effective resistance of the two parallel resistors $= 4$ Ω [1].
Circuit current, $I = V/R = 3.0$ V $\div$ 7.5 Ω [1] $= 0.40$ A [1].

(b) $V = IR = 0.40$ A $\times 7.0$ Ω [1] $= 2.8$ V [1]. [Total: 5]

Chapter 3 Waves, imaging and information

1 (a) In a transverse wave the vibrations are at right angles to the direction of wave travel [1].
Examples include any electromagnetic wave, e.g. radio, light [1].
In a longitudinal wave the vibrations are parallel to the direction of wave travel [1].
Examples include sound or any other compression wave [1].

(b) (i) In an unpolarised wave the vibrations are in all directions perpendicular to the direction of travel [1]. In a polarised wave the vibrations are in one direction only [1].

(ii) Sound is a longitudinal wave; there are no vibrations perpendicular to the direction of travel [1]. [Total: 7]

2 (a) The incident light, refracted light and normal are all in the same plane [1].
$\sin i \div \sin r =$ constant [1].

(b) (i) $\lambda = v \div f$ [1] $= 3.00 \times 10^8$ m s^{-1} $\div$ 6.00×10^{14} Hz [1] $= 5.00 \times 10^{-7}$ m [1]

(ii) $n = c_{air} \div c_{water} = 3.00 \times 10^8$ m s^{-1} $\div$ 2.25×10^8 m s^{-1} [1] $= 1.33$ [1]

(iii) $\sin r = \sin i \div n = \sin 63° \div 1.33$ [1] $r = \sin^{-1} 0.668 = 42°$ [1]. [Total: 9]

3 (a) (i) 1.2 m [1]

(ii) $f = 1/T = 1 \div 2.50 \times 10^{-4}$ s [1] $= 4000$ Hz [1]

(iii) $v = f\lambda$ [1] $= 2000$ Hz $\times 1.2$ m [1] $= 2400$ m s^{-1} [1].

(b) (i) The wave profile does not move along the rope [1]

(ii) By superposition [1] of a wave and its reflection [1].

(c) (i) $0°$ [1]

(ii) $180°$ [1]. [Total: 11]

4 (a) (i) The waves are in phase [1] and interfere constructively [1]

(ii) The waves are out of phase [1] and interfere destructively [1].

(b) (i) $\lambda = v \div f$ [1] $= 330$ m s^{-1} $\div$ 2000 Hz $= 0.165$ m [1]

(ii) Separation of maxima, $x = \lambda D/a = 0.165$ m $\times 5.0$ m $\div$ 0.75 m $= 1.10$ m [1].
CE is half this distance $= 0.55$ m [1]. [Total: 8]

5 (a) $_gn_p = _an_p \div _an_g$ [1] $= 1.24 \div 1.58 = 0.785$ [1].

(b) $\sin C = 0.785$ [1] $C = \sin^{-1} 0.785 = 52°$ [1].

(c) Light has further to travel along path B than along path A [1]. This causes a pulse to be elongated [1]. [Total: 6]

Chapter 4 Waves, particles and the Universe

1 (a) (i) Particles that approach a nucleus are deflected away from it [1]. So the nucleus must have the same sign of charge as the alpha particles [1]
    (ii) Many particles pass through undeviated [1].
(b) (i) High energy electrons [1] are used to penetrate nucleons [1]
    (ii) Electrons are not repelled by the nucleus [1] and they are not affected by the strong nuclear force [1]
    (iii) The proton and neutron are made up of other particles [1]
    (iv) The scattering of the electrons shows that there are dense regions of charge within a nucleon [1]
    (v) A proton is uud [1]; a neutron is udd [1]. [Total: 11]

2 (a) 6.02×10^{23} mol^{-1} × 1/12 mol [1] = 5.02×10^{22} atoms [1].
(b) (i) No [1] since 99% of carbon is carbon-12 [1]
    (ii) 5×10^{10} [1]. [Total: 5]

3 (a) If the neutron–proton ratio is less than 1 the nucleus undergoes β^+ decay [1]; if the neutron–proton ratio is greater than 1 the nucleus decays by β^- emission [1].
(b) When $^{13}_{6}$C decays, the neutron–proton ratio changes from 1.17 [1] to 0.86, which is closer to 1 [1]
(c) It becomes greater than 1 [1] and reaches a value of 1.5 for massive nuclei [1]. [Total: 6]

4 (a) A meson consists of a quark and an antiquark [1]. A baryon is made up of three quarks [1].
(b) (i) π^0 [1].
    (ii) Other pions have charge [1]. Exchange particle here must be neutral [1]. [Total: 5]

5 (a) (i) Emission of electrons [1] when a metal surface is illuminated with electromagnetic radiation [1]
    (ii) A quantum of electromagnetic radiation [1] of a particular frequency [1]
    (iii) The minimum energy [1] an electron needs to escape from a metal [1].
(b) (i) $f_0 = \phi \div h$ [1] = 5.05 eV × 1.60×10^{-19} C ÷ 6.63×10^{-34} J s = 1.22×10^{15} Hz [1]
    (ii) $(\tfrac{1}{2}mv^2)_{max} = hf - \phi$ [1] = 6.63×10^{-34} J s × 1.80×10^{15} Hz – (5.05 eV × 1.60×10^{-19} C) [1] = 3.85×10^{-19} J [1]. [Total: 11]

Index